Burleigh

INTEGRATED PEST MANAGEMENT FOR

ALMONDS

UNIVERSITY OF CALIFORNIA
STATEWIDE INTEGRATED PEST MANAGEMENT PROJECT
DIVISION OF AGRICULTURE AND NATURAL RESOURCES
PUBLICATION 3308

1985

PRECAUTIONS FOR USING PESTICIDES

Pesticides are poisonous and must be used with caution. READ THE LABEL BEFORE OPENING A PESTICIDE CONTAINER. Follow all label precautions and directions, including requirements for protective equipment. Use a pesticide only against pests specified on the label or in published University of California recommendations. Apply pesticides at the rates specified on the label or at lower rates if suggested in this publication. Laws, regulations, and information concerning pesticides change frequently, so be sure the publication you are using is up to date.

Legal Responsibility. The user is legally responsible for any damage due to misuse of pesticides. Responsibility extends to effects caused by drift, runoff, or residues.

Transportation. Do not ship or carry pesticides together with food or feed in a way that allows contamination of the edible items. Never transport pesticides in a closed passenger vehicle or in a closed cab.

Storage. Keep pesticides in original containers until used. Store them in a locked cabinet, building, or fenced area where they are not accessible to children, unauthorized persons, pets, or livestock. DO NOT store pesticides with foods, feed, fertilizers, or other materials that may become contaminated by the pesticides.

Container Disposal. Dispose of empty containers carefully. Never reuse them. Make sure empty containers are not accessible to children or animals. Never dispose of containers where they may contaminate water supplies or natural waterways. Consult your county agricultural commissioner for correct procedures for handling and disposal of large quantities of empty containers.

Protection of Non-Pest Animals and Plants. Many pesticides are toxic to useful or desirable animals, including honeybees, natural fish, domestic animals, and birds. Crops and other plants may also be damaged by misapplied pesticides. Take precautions to protect non-pest species from direct exposure to pesticides and from contamination due to drift, runoff, or residues. Certain rodenticides may pose a special hazard to animals that eat poisoned rodents.

Posting Treated Fields. For some materials, re-entry intervals are established to protect field workers. Keep workers out of the field for the required time after application and, when required by regulations, post the treated areas with signs indicating the safe re-entry date.

Harvest Intervals. Some material or rates cannot be used in certain crops within a specified time before harvest. Follow pesticide label instructions and allow the required time between application and harvest.

Permit Requirements. Many pesticides require a permit from the county agricultural commissioner before possession or use. When such materials are recommended in this publication, they are marked with an asterisk (*).

Processed Crops. Some processors will not accept a crop treated with certain chemicals. If your crop is going to a processor, be sure to check with the processor before applying a pesticide.

Crop Injury. Certain chemicals may cause injury to crops (phytotoxicity) under certain conditions. Always consult the label for limitations. Before applying any pesticide, take into account the stage of plant development, the soil type and condition, the temperature, moisture, and wind. Injury may also result from the use of incompatible materials.

Personal Safety. Follow label directions carefully. Avoid splashing, spilling, leaks, spray drift, and contamination of clothing. NEVER eat, smoke, drink, chew while using pesticides. Provide for emergency medical care IN ADVANCE as required by regulation.

To simplify information, trade names of products have been used. No endorsement of named products is intended, nor is criticism implied of similar products which are not mentioned.

ORDERING

Single copies of this publication, #3308, may be ordered by sending check or money order payable to The Regents of the University of California to:

Publications
Division of Agriculture and Natural Resources
University of California
6701 San Pablo Avenue
Oakland, CA 94608–1239
(415) 642-2431

Information on price, foreign orders, and quantity discounts may be obtained by contacting the above office.

ISBN: 0–931876–73–7

Library of Congress No. 85–70728

6m–11/85

Printed in the United States of America

Contents

Contributors and Acknowledgments

This manual was produced under the auspices of the Statewide IPM Project, James M. Lyons, Director, Howard Ferris, Associate Director.

Prepared by the IPM Manual Group of the Statewide IPM Project, U.C. Davis, Mary Louise Flint, Director.

Barbara Peterson Ohlendorf, Senior Writer
Jack Kelly Clark, Principal Photographer

Technical Coordinators

Martin M. Barnes
Professor of Entomology, U.C. Riverside

William W. Barnett
Area IPM Specialist, Cooperative Extension, Fresno

Dale E. Kester
Professor of Pomology, U.C. Davis

Warren C. Micke
Extension Pomologist, Cooperative Extension, U.C. Davis

Beth L. Teviotdale
Plant Pathologist, Cooperative Extension, Kearney Agricultural Center

Contributors

Entomology: Martin M. Barnes, William W. Barnett, Walter J. Bentley, Robert K. Curtis, Marjorie A. Hoy, Richard E. Rice, Craig V. Weakley, Frank G. Zalom

Nematology: Michael V. McKenry

Pesticide Application: John E. Dibble

Plant Pathology: Richard M. Bostock, W. Harley English, John M. Mircetich, George Nyland, Joseph M. Ogawa, Beth L. Teviotdale

Pomology: Joseph H. Connell, Dale E. Kester, Warren C. Micke, G. Steven Sibbett, Mario Viveros

Vertebrates: Terrell P. Salmon

Water Management: David A. Goldhamer, Terry L. Prichard, Richard L. Snyder

Weed Science: Clyde L. Elmore, Bill B. Fischer, Art H. Lange

Special Thanks To:

T.M. Aldrich, W.K. Asai, R.F. Brewer, L.E. Caltagirone, J.P. Clark, W.R. Clark, C.E. Curtis, T.D. Eichlin, B.R. Hanson, J.K. Hasey, J.C. Headley, L.C. Hendricks, E. Kalar, C. Kitayama, B. Kobbe, W.H. Krueger, J.M. Labavitch, S.E. Lindow, R.E. Marsh, R.D. Meyer, A.H. Purcell, W.O. Reil, D. Rough, P.A. Rude, R.R. Sanborn, R. Sanders, J.P. Sanderson, W. C. Schnathorst, M.W. Stimmann, L.L. Strand, R.W. Thorp, K. Uriu, R.A. Van Steenwyk, R.N. Vargas, S.A. Weinbaum, B. Wilk, L.T. Wilson, R.R. Youngman, M.L. Zavala for suggestions, manuscript review, or help with photographs.

Information in this manual has been derived from research supported by the University of California, the USDA, and the Almond Board. Support for the compilation of this manual was provided by the Almond Board of California and the University of California Statewide IPM Project. Thanks to Agriculture and Natural Resources Publications, University of California, for making available editors and artists.

Production

Manuscript Preparation: Betty Rudd
Editing: Peggy Anne Davis
Drawings: Marvin Ehrlich
Design and Production Coordination: Naomi Schiff

Figure 1. Almond-growing areas in California (one dot equals 1,000 acres). (Calif. Crop & Livestock Reporting Service)

Integrated Pest Management for Almonds

Integrated Pest Management (IPM) provides a framework for making pest control decisions that minimize losses by pests with as little cost to the grower and disruption of the environment as possible. This manual presents IPM methods and strategies for almond crop production and is intended for use by anyone interested in almond production in California.

California is the primary place in North America where almonds are grown commercially. A small percentage of the almond acreage in California is located in coastal districts such as Contra Costa and San Luis Obispo counties, but the majority of the state's 425,000 acres of almonds is in the Central Valley (Figure 1). Although orchards of all ages occur in all districts, many orchards located in the coastal districts are quite old and some are unirrigated. Almond orchards in the central and northern portions of the Central Valley are usually flood, drip, or sprinkler irrigated. These orchards, on the average, are older and smaller than those located in the southern portion of the Central Valley. Orchards in the southern Central Valley are usually more closely planted than some of the older orchards in other districts, often in blocks of several hundred acres or more, and are either flood, drip, or sprinkler irrigated.

Major almond cultivars (varieties) are self-sterile; the trees must be pollinated with pollen from another almond cultivar. For this reason, more than one cultivar is planted in an orchard, and the trees are cross-pollinated by honey bees.

Almond production has been more successful in California's Central Valley than in other regions because the Mediterranean-type climate in the Central Valley is favorable to the crop's exacting requirements. Although almonds require winter chilling to break dormancy, this requirement is less than for most other deciduous fruit and nut crops. Almonds need considerable heat during the growing season; thus almond trees usually produce best in the warm, dry interior valleys. Almonds bloom in February to early March, and from blossom time through April, the young nuts are susceptible to frost damage. Almond trees are also susceptible to a variety of diseases during the bloom period, so ideally rain should be minimal at this time and the days warm enough (55° to 60°F or higher; 13° to 15.5°C) for the bees to pollinate the trees. Rain at any time during the growing season can cause serious disease problems, and late in the season it can interfere with harvest operations. This is why a Mediterranean-type climate with a rainy, mild winter followed by a warm, rainless spring and summer is ideal for almond growing.

By mid- to late March the trees have leafed out and small nuts can be seen. Nuts grow in size up to early May, when the embryo begins to appear. The embryo develops into the kernel during May and early June, and nut maturation begins in July when hulls begin to split open. As the split widens, the hull separates from the shell. Hulls and kernels may dry either on the tree or on the ground after the nuts have been knocked from the trees. Almonds are ready to knock from the tree when 95 to 100% of the hulls have begun to split open. This generally occurs from early August to late September, depending on the cultivar and the location in the state where it is grown.

Disease management is very important during the spring, especially at bloom, because of possible rain. Other management activities at this time include frost protection, crop pollination, surveying weeds, and monitoring vertebrate and insect populations.

The busiest period of management activities in the orchard occurs during the warm part of the growing season. Activities include irrigation, weed management, sampling nutrient levels, and monitoring for insect, mite, disease, and vertebrate problems. Nuts remaining on the tree after harvest should be removed because they provide overwintering sites for navel orangeworm, as well as brown rot fungi. Winter orchard management operations include pruning, applying dormant sprays and preemergence herbicides, and monitoring for perennial weed seedlings.

This manual contains separate chapters on the five major categories of orchard pests: insects and mites, diseases, weeds, nematodes, and vertebrates. Introductory sections of these chapters explain where and when major pests occur and general monitoring guidelines. Each chapter gives detailed descriptions and photographs of individual pests, the damage they cause, and specific guidelines for enhancing natural control factors, designing a monitoring program, and using control actions effectively.

Introductory chapters on almond growth and development and general management practices provide the background upon which management guidelines in the pest sections are based. This manual is intended to be used in conjunction with *Almond Orchard Management*, also published by the University of California Division of Agriculture and Natural Resources, and listed in the reference section of this publication. *Almond Orchard Management* has more in-depth information on cultural methods and specific crop requirements than this publication does.

As knowledge about specific pests increases, and the interaction between these pests and the orchard environment is better understood, monitoring and control techniques will be modified and updated. Check regularly with your farm advisor for new developments. Because pesticide registrations change frequently, few specific recommendations are given in this manual. For pesticide recommendations, see *A Guide to Controlling Almond Pests, Diseases, and Micronutrient Deficiencies*, listed under Pesticides in References.

The Almond Tree: Development and Growth Requirements

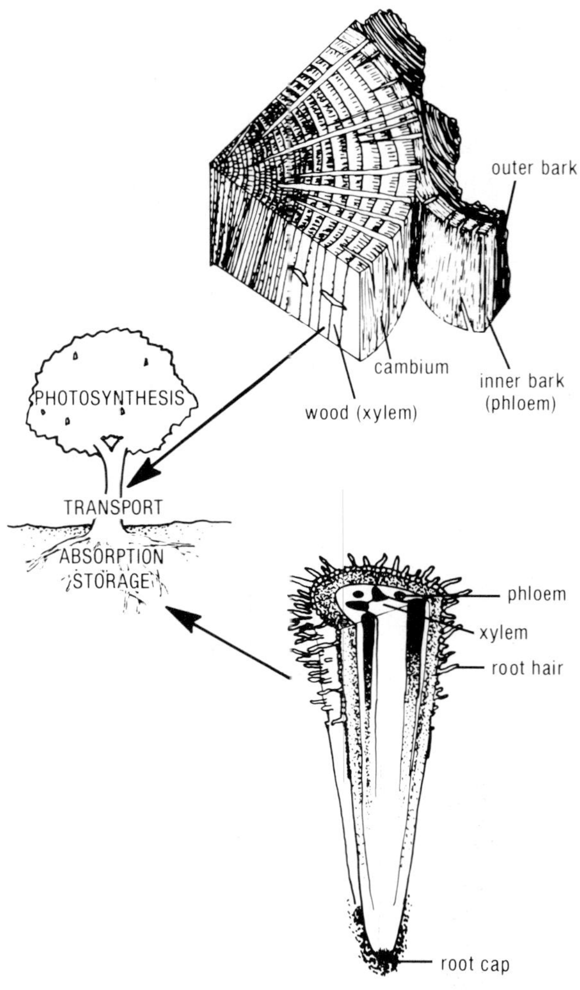

Figure 2. Basic tree structures and functions. Young roots absorb water and elements, which are transported in the xylem to the canopy, where leaves produce sugars in the process of photosynthesis. The tree uses sugars immediately for growth or transports them in the phloem mainly to the roots for storage.

The objective of almond production is to harvest a large crop of high quality kernels in the most cost-effective manner. Quantity or total yield of a crop is determined by number of nuts produced and size and weight of individual nuts. Each yield component is affected by different factors in the orchard environment, including pest management. The number of nuts produced on a tree, for example, depends on the number of flowers, the percentage set, and the proportion that matures. Because flowers are borne laterally on spurs and on shoots, adequate growth from terminal buds on spurs or growth of lateral shoots is essential to maintain high yields in consecutive years. Individual kernel size and weight depend on the cultivar, number of kernels per unit area, and management practices.

Optimal almond production requires healthy trees. An orchard site must be properly prepared before planting to ensure that the trees get off to a good start. New trees require special irrigation, training, and pest control procedures during the nonbearing years while they are becoming established. Maintaining tree health once the orchard is established requires a basic understanding of the tree's seasonal cycle, the crop's specific requirements, and the impact that pest organisms and cultural and environmental factors have on tree and nut growth. Without this understanding, it is easy to overlook stress symptoms caused by the environment or to confuse them with pest damage. An important aspect of an IPM program is the evaluation of cultural practices and pest control methods for their impact on the total orchard system. In some cases, well intended practices could compound tree stress and add to economic loss.

The Nonbearing Years

Almond trees are classified by the California Crop and Livestock Reporting Service (CCLRS) as nonbearing until the fifth growing season (fifth leaf stage), or 4 years after planting, although some growers may harvest in the fourth or even third growing season. During the nonbearing years, the major period of root growth occurs and the

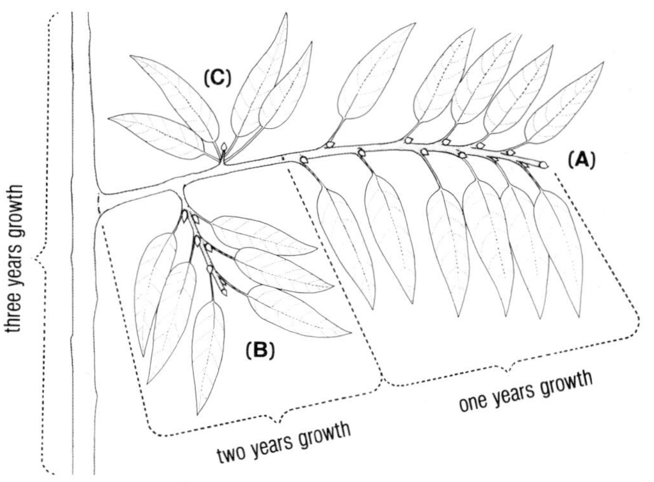

Figure 3. Schematic drawing of the three locations of vegetative growth: (A) From terminal bud of a long shoot (more than 25 nodes), lateral buds may be either vegetative or flower; if vegetative they can produce shoot growth. (B) From terminal bud of elongated spur or short shoot (8 to 25 nodes), lateral buds mostly flower, sometimes vegetative. (C) From terminal bud of a spur (less than 8 nodes), lateral buds are flowers.

basic framework of the tree is developed. These years are critical to the future well-being and productivity of the tree.

The first year that the tree is in the ground is the most important for root development. Stress caused by disease, nematodes, weed competition, or insufficient irrigation greatly hinders root development and hence top development, even in trees planted on resistant rootstocks. After the first few years, young trees become more tolerant to stress.

The tree undergoes two to three growth flushes during its first year in the orchard. Like root development, however, the amount of top growth depends on weed competition, the availability of water, and other cultural and environmental factors. These growth flushes occur during spring and summer. By late summer or when water is insufficient, growth slows down.

Following the first growing season, the tree is pruned (trained) to select the primary scaffold limbs. Shoot strikes caused by the peach twig borer or oriental fruit moth can make it difficult to shape the tree at this time. Shoot strikes reduce the amount of terminal growth on branches and cause lateral buds to develop and grow. Excess lateral branch growth reduces the number of good branches to choose from when selecting the primary scaffolds.

Tree growth during the second year also occurs in growth flushes, but the major growing period may be reduced. Most above-ground growth occurs between March and mid-May. Below-ground growth of the roots takes place primarily in summer and fall. Another important training period follows the second growing season. At this time, the secondary structure of the tree is selected, thus establishing its basic framework. Shoot strikes in the spring and early summer of the second year will also create problems by reducing the amount of terminal growth and, therefore, the number of good branches to choose from.

If the tree has adequate irrigation and other growing conditions are not limiting, flower buds are initiated on long shoots during the second or third growing season. These flower buds bloom the following season, and fruiting spurs then begin to develop on lateral shoots. If adequate shoot and spur development occurs during the second and third growing seasons, the tree can become productive by the fourth year. Spurs continue to develop laterally on these long shoots and bear fruit in subsequent growing seasons.

The Seasonal Cycle of Bearing Trees

In November, the almond tree enters a period of rest that lasts through December. During this time, the tree maintains a base level of water transport and starch consumption. In winter, the breakdown of starch increases sugar concentration in the cell sap, preventing it from freezing. To produce enough starch for this process, the tree must have been provided with sufficient water in the fall.

Developing vegetative and flower buds require adequate moisture and exposure to a certain amount of chilling in fall and winter to overcome the rest period. Usually by January 1, the chilling requirement has been met and the buds begin to expand. At this time, the tree's water and nutrient conducting system, the xylem and phloem, and the storage areas are loaded with carbohydrates, nutrients, amino acids, and hormones, all of which are essential for growth (Figure 2).

Flower buds usually begin expanding somewhat ahead of leaf and shoot buds, but this may vary with cultivar and season. Shoot growth begins from mid-February to mid-March—the time of flowering—and occurs at three locations on the twig (Figure 3): from the terminal bud of a spur, from the terminal bud of a shoot, and from lateral buds of 1-year-old shoots. As leaves unfold and expand, photosynthetic production of carbohydrates starts, and growth is no longer entirely dependent on carbohydrates stored from the previous year.

The eight flowering stages for almonds are pictured on the back cover of this manual. Bloom time for a particular cultivar is important because almond trees need

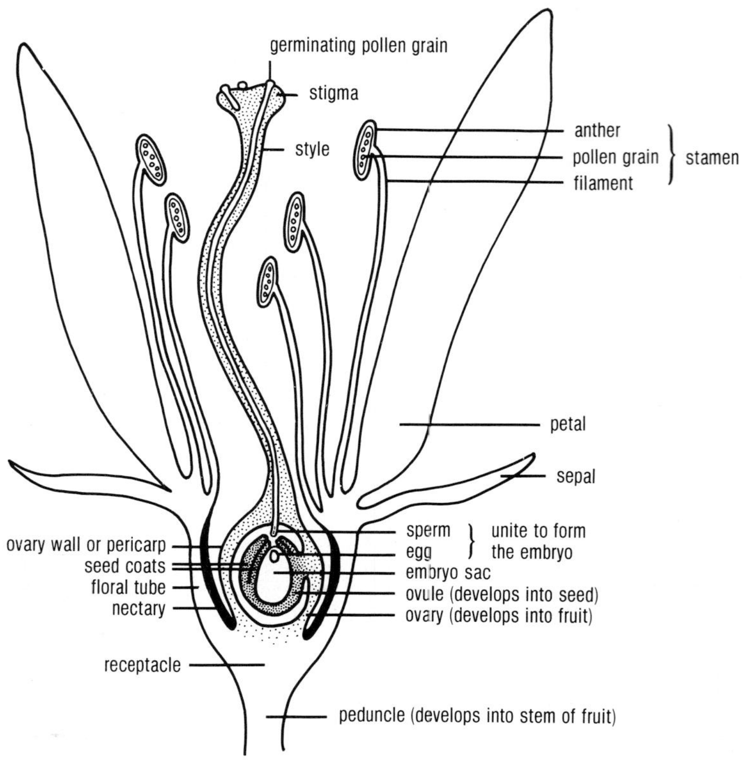

Figure 4. Diagrammatic drawing of a longitudinal section of an almond flower. Pollination has occurred. The pollen tube formed by subsequent germination of a pollen grain carries two sperms to the embryo sac, where one of them unites with the egg (fertilization). This union results in an embryo and development of the ovary into a fruit. (Drawing by Reid M. Brooks.)

cross-pollination. The amount of winter chilling that trees receive greatly affects the degree to which the bloom periods for different cultivars overlap. In general, insufficient winter chilling results in a delayed bloom that is prolonged and straggly. Late blooming cultivars are affected more severely by insufficient winter chilling than are early blooming cultivars.

Pollination occurs when one or more grains of pollen are transferred from anthers of one cultivar's flower to the stigma of another cultivar's flower (Figure 4). This work is done by insects; in commercial orchards, it is essentially all done by honey bees. It is critical for adequate crop set that enough bees are present in the orchard at the right time. The job is not complete, however, when pollination takes place. To produce a harvestable nut, the pollen must germinate to form a pollen tube that grows down the style of the almond flower. Sperm are then released into the flower's embryo sac at the base of the style to fertilize the egg.

Weather is important in the pollination (setting) and fertilization of the crop. Bees are generally not active when temperatures are below 55°F (13°C), in rainy weather, or when winds exceed 15 miles per hour. Temperatures at which bees are active closely coincide with the temperature range that is best for growth of the pollen tube. If temperatures are too low, the pollen tube grows so slowly that the embryo sac may lose its viability before fertilization can take place. For more information, refer to *Almond Orchard Management* and *Honey Bees in Almond Pollination*, both listed in References.

A period of rapid hull and shell growth follows bloom and continues until early May. During this stage of fruit growth, potential size of the nut (length and width) is established. The more nuts on a spur the smaller the average nut size. Some of the small nuts will cease development and eventually drop. This loss of nuts is a natural thinning process that probably results from competition among too many nuts.

During hull and shell growth in April, the available energy of the tree is directed toward both shoot and nut growth. These two processes compete directly with each other for nutrients and water during this period and can both be reduced by either moisture stress or nutrient deficiencies.

As the nut reaches full size in late April, the inner layer of the shell begins to harden, while the outer portion remains soft. This is a critical period with some cultivars because growth stress at this time can cause splits or cracks in the inner shell, while the outer shell, which is still soft, is capable of producing callus tissue. Cracks and splits may disrupt the conducting tissues in the nut and cause embryo abortion, gumming in the hull, or corky layers. All of these conditions can lead to kernel loss or reduction in quality.

The embryo or kernel begins to enlarge rapidly in early May. This process lasts until early June. Initially, the internal tissue of the embryo sac (nucellus) is clear and watery. Then a distinct translucent tissue, the endosperm, becomes evident at the tip of the nucellus. Within a few weeks, a tiny white embryo (kernel) develops within the endosperm (Figure 5). Moisture stress during the period when the embryo is growing could result in an aborted or partially filled and shriveled kernel.

By early June, the kernel is formed, and dry weight accumulation begins in the kernel and continues as the nut matures. Dry weight depends on carbohydrates produced by photosynthesis in leaves. Therefore, good foliage and adequate moisture contribute to final kernel weight. Likewise, defoliation and moisture stress during the period when the kernel is increasing in weight can adversely affect quality as well as final kernel weight and, therefore, yield. Hardening of the shell takes place in June and July.

The approach of nut maturity is signaled by the initiation of hull split—that is, the separation (dehiscence) of the hull from the shell. The process of hull split is defined in six stages (Figure 6, page 12). The first stage is the unsplit nut. Stage 2 is divided into three parts and begins with the formation of a thin separation line at the suture depression that separates the two sides of the hull (Stage 2a). This suture line gradually deepens and extends from tip to base, but the hull cannot be squeezed open at the suture (Stage 2b). In the final part of Stage 2, the hull develops a separation layer. Although an opening may not be visible, the hull can be squeezed open by pressing both ends of the hull (Stage 2c). Soon the opening becomes visible (Stage 3), and the hull separates completely to expose the nut (Stage 4).

At the same time that hull split is occurring, an abscission layer is being formed where the nut attaches to the peduncle (fruit stem). Once this layer is completely developed, the nut remains attached to the tree only by short fibers, which can be broken when the nuts are shaken or knocked.

As the abscission layer forms and the nut is exposed, the edge of the hull begins to dehydrate (Stage 5). With exposure and dehydration, the shell and the pellicle (skin of the kernel) change from white to brown. If the nuts are removed from the tree, dehydration and browning will continue on the ground until eventually, the entire hull and nut are dried (Stage 6).

Hull split requires adequate tree moisture to properly proceed because the hull must be turgid. Moisture stress can change the sequence of ripening, triggering early abscission and hull split. These early splitting hulls usually open improperly, or tighten on the shell, and become sticktights. On the other hand, hull split and abscission layer formation may be delayed in trees maintained under excessively vigorous conditions.

When hull split occurs is important in a pest management program because it exposes nuts to invasion by navel orangeworm and hull rot fungi. The longer the nuts remain on the tree after hull split, the more vulnerable they are. The exact time of hull split in an orchard is complicated, however, because ripening does not occur in all nuts on a tree at the same time. It begins in the upper and outer periphery of the tree and later extends to the lower and inner sections. The process of nut ripening throughout the tree may take 3 to 4 weeks.

Almonds should be harvested when 95 to 100% of the nuts have split and adequate nut removal can be achieved (Stage 3 to 4). The start of almond harvest varies from year to year and district to district, but it usually begins in early to mid-August. Many of the nuts will have progressed to Stage 4 or more, but some will still have hulls that are moist and green. These will dry rapidly on the ground however. It is important to harvest at the earliest possible date to minimize navel orangeworm damage to nuts and to avoid complications caused by early rains. If the threat of a navel orangeworm infestation is severe, orchards can be harvested twice: once to remove

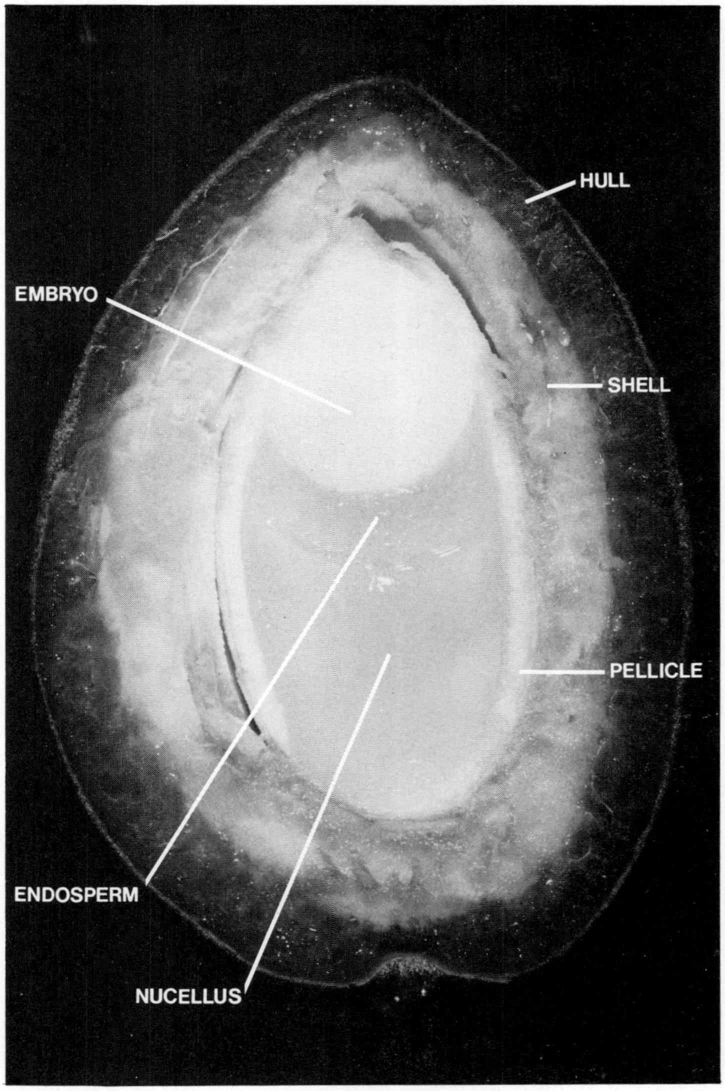

Figure 5. Cross section of an immature nut shows developing embryo and endosperm.

the early ripening nuts and a second time to remove the later ripening ones.

While new nuts are developing in late spring and summer, the tree is also initiating buds for next year's crop. As a result, stress during one growing season also can have an impact on the next season's yields. Bud initiation occurs mainly during April and the early part of May. Buds then enter a period of dormancy that lasts from June through mid-August, although they may be forced out of dormancy prematurely by severe water stress or defoliation followed by irrigation. In mid-August, some of the buds in the leaf axils begin to differentiate into flower buds, while others remain as vegetative buds. About October, depending on cultivar, buds enter

1

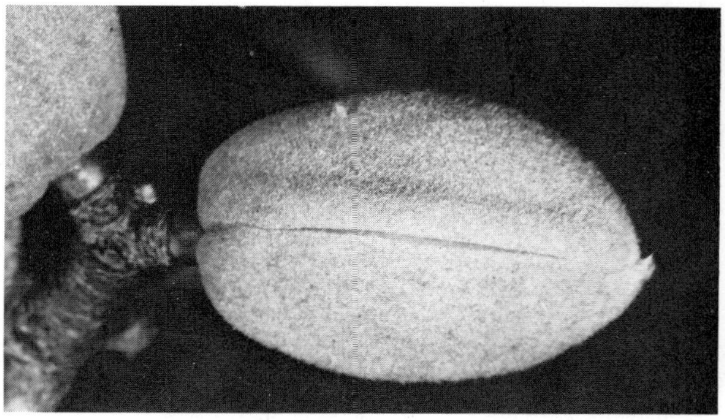

2a

2b

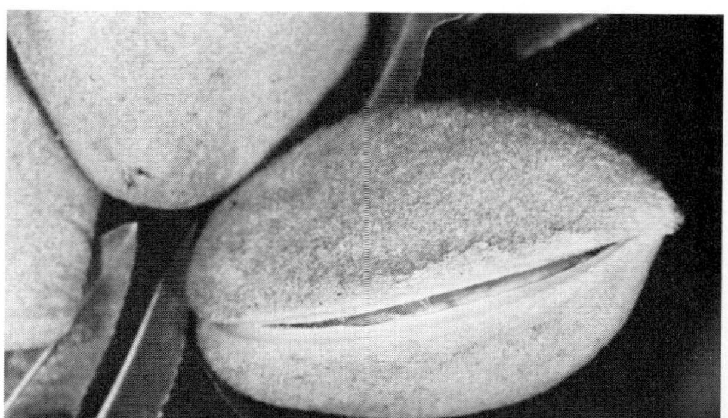

2c

3

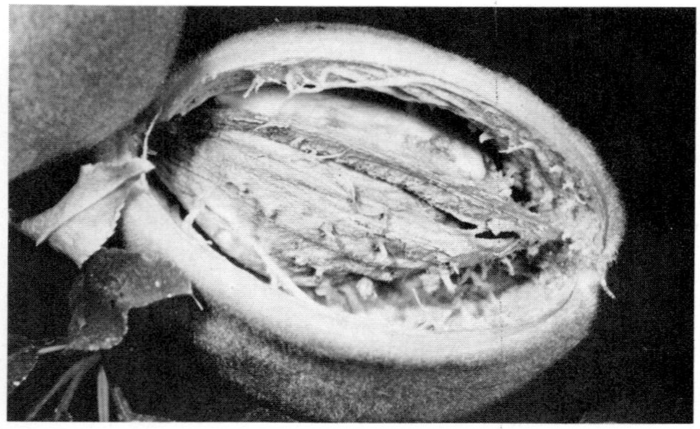

4

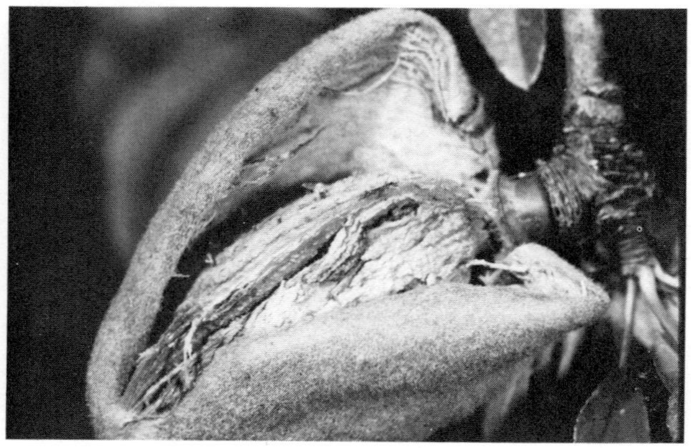

5

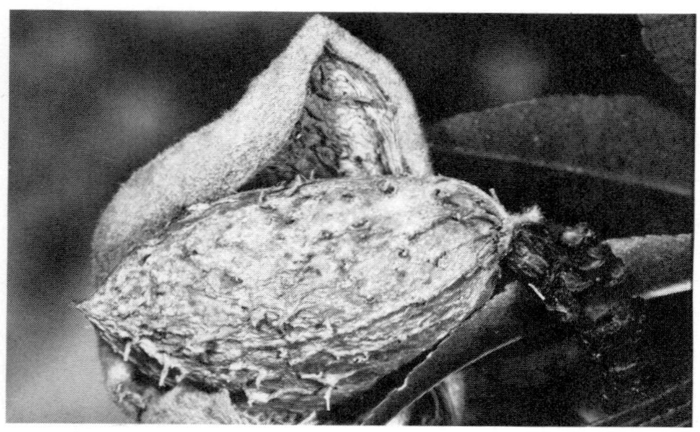

6

a rest period that lasts through December; neither vegetative nor flower buds can be forced into growth during this period. A certain amount of chilling during this resting period is required for normal flowering.

During the growing season, sugar for the tree is synthesized in the leaves. By fall, the tree begins to store much of this sugar as starch to be used for next year's growth flush. The tree also begins preparation for winter by decreasing its total moisture content and synthesizing proteins needed for the rest period.

Growth Requirements

Tree growth and nut development require sufficient light, water, and nutrients, suitable soil, and adequate temperatures. Light is essential for photosynthesis, the process by which green plants manufacture sugars, their major food source (Figure 7). Sugars are synthesized in the leaves from water taken from the soil, carbon dioxide taken from the air, and light energy from the sun. Sugar provides energy for tree growth, development, and maintenance and may be used immediately or stored as starch in leaves, stems, or roots.

Irrigation is necessary for profitable almond production in the semi-arid climate of California. In the Central Valley, a mature almond orchard requires 30 to 48 acre-inches of applied irrigation water per year, depending on rainfall, water use, and irrigation efficiency, for optimal yield of high quality nuts. Only a small fraction of the water taken up by the tree is retained; most of it passes through the plant and out the leaf pores during transpiration. Transpiration is necessary to supply water to the leaves for photosynthesis, to carry nutrients to all parts of the tree, and to cool the plant. Climatic factors, such as the amount of sunshine, humidity, heat and wind, affect the rate of transpiration.

Trees need various nutrients for growth and nut production. California orchard soils contain sufficient amounts of most essential nutrients, either because they are already present in the soil or are returned periodically through leaf decay. Others, such as nitrogen and zinc, have to be replenished through fertilization.

Almond trees grow best on deep, well-drained, non-stratified, light- to medium-textured soils. Under ideal soil conditions, roots may grow to depths of 9 feet (3 m) or more. Soil conditions, physical or chemical, that limit the rooting system or affect the health of roots have a direct influence on the trees' size, vigor, and crop potential.

Almond trees also grow best under warm temperatures but require a certain amount of winter chilling for uniform bloom and good fruit set in the spring. Almonds are susceptible to extreme temperatures, which can cause frost damage to blossoms or young nuts in the spring or heat stress in summer.

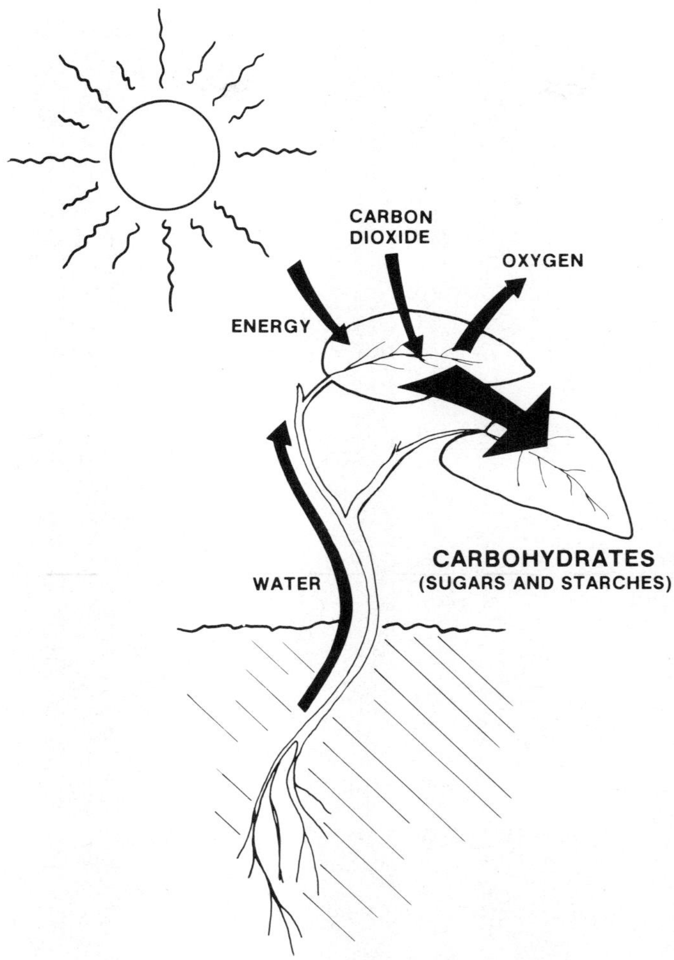

Figure 7. In photosynthesis, green plants use energy from the sun to convert carbon dioxide from the air and water from the soil into sugars, their primary food source. In the process, oxygen is released into the air.

◀ Figure 6. Stages of hull split are as follows: 1) unsplit nut—no evidence of a separation at the suture; 2a) initial separation—50% or more of a thin separation line visible; 2b) deep V, unsplit—beginning of splitting with a deep "V" over at least 50% of the suture line, but the hull cannot be squeezed open at the suture; 2c) deep V, split—a deep "V" in the suture, which is not yet visibly separated, but which can be squeezed open by pressing both ends of the hull; 3) split, less than 3/8 inch (1 cm)—a visible opening of the suture less than 3/8 inch at midsuture; 4) split, more than 3/8 inch—a visible opening of the suture greater than 3/8 inch at midsuture; 5) initial drying—the edges of the hull are beginning to dry; 6) completely dry.

Managing Pests in Almonds

Integrated pest management treats pests as part of a crop production system that includes not only the crop and its pests, but also the physical and biological environment in which the crop is grown. A good IPM program coordinates pest management activities with cultural operations to achieve economical and long-lasting solutions to pest problems. The emphasis is on using monitoring techniques for optimum timing of control, developing a good prevention program, and selecting the best available control method based on knowledge of the pest's life history. Minimizing pest-caused crop losses and environmental damage in the most cost effective manner are the ultimate goals of IPM. Figure 8 lists the cultural operations and pest management actions most likely to be used during the year. This summary sheet also gives page numbers in this manual where details about a particular management practice can be found.

An integrated pest management program has four major components:

- pest identification
- field monitoring
- control action guidelines
- management methods for prevention and control

Pest Identification

Correctly identifying pest species and stress symptoms is fundamental to choosing the most appropriate management strategy. Most pest management tools, including pesticides, are effective only against a select group of organisms; therefore, different practices may be needed even for closely related organisms. In some cases, symptoms caused by pest organisms closely resemble those caused by nutrient deficiencies or soil problems. The photographs, descriptions, and seasonal charts in this manual are intended to help identify problems. Farm advisors or pest control advisers (PCAs) can assist you in confirming identification or in identifying unusual problems.

Field Monitoring

Field monitoring provides information on daily or seasonal orchard conditions, such as the status of pests, weather, or soil moisture. This information can be used to evaluate and predict potential pest or orchard problems. Each orchard must be monitored because conditions vary between neighboring orchards.

Regularly check pest and beneficial species present, maturity and health of the crop, weather, and soil moisture. Keep records of monitoring results, weather, and management activities. During the season, these records can show whether pests or natural enemy populations are rising or decreasing. They can also help forecast possible outbreaks of the next generation of a pest. Pest population counts, together with records of control measures, cultural practices and weather conditions, can help you decide whether future control actions will be needed. Making simple tables and graphs from the data you collect will help you identify population patterns. Mapping soil or weed problems is helpful in identifying localized problems and pest distribution. Over the years,

MANAGEMENT ACTION	Page
WINTER (Nov-Jan)	
☐ Remove mummy nuts from trees and destroy	30,58
☐ Prune trees	30
☐ Apply dormant spray for peach twig borer, San Jose scale, and mite eggs	53
☐ Monitor winter annual and perennial weed seedlings (Use record sheet, Figure 59)	141
☐ Monitor pocket gophers, control if needed	43
☐ Arrange to have beehives placed in orchard prior to bloom	9-10

MANAGEMENT ACTION	Page
SPRING (Feb-Apr)	
☐ Protect against frost as needed	23
☐ Mow ground cover prior to bloom and thereafter as needed	29,136
☐ Bloom and Post Bloom Disease Control	
1) Protect against brown rot as needed	108
2) Protect against scab, shot hole, and leaf blight as needed	111-114
3) Protect against green fruit rot, if needed	114
☐ Begin irrigation as needed	24-27
☐ Monitor winter annual and perennial weeds (Use record sheet, Figure 59)	141
☐ Apply postemergence herbicides as needed	139
☐ Apply nitrogen fertilizer in April (first half of a split treatment)	27
☐ Begin Monitoring Insect and Mite Pests	
1) Set out egg traps for navel orangeworm, control if needed	60
2) Set out pheromone traps for peach twig borer and calculate degree-days, control if needed	68
3) Set out pheromone or sticky tape traps for San Jose scale if needed and calculate degree-days	71
4) Set out pheromone traps for oriental fruit moth, if needed	85
5) Monitor European red and brown mites, control if needed	81,83
6) Make an initial survey for ants	84
7) Monitor for American plum borer, if needed	87
☐ Monitor vertebrates (ground squirrels, pocket gophers), control if needed	39,43

MANAGEMENT ACTION	Page
SUMMER (May-Jul)	
☐ Irrigate as needed	24-27
☐ Monitor summer annual weed seedlings and perennials (Use record sheet, Figure 59)	141
☐ Continue Monitoring Insects and Mites	
1) Monitoring webspinning and European red mites	78
2) Monitor ants after an irrigation	84
3) Monitor for American plum borer and other wood-boring insects if needed	86-89
☐ Have leaf analysis done (July)	29
☐ Monitor vertebrate (ground squirrels, birds, gophers) activity, control if needed	39,43 43
☐ Monitor hull split for spray and earliest harvest date, and apply sprays for navel orangeworm if needed	10-12 64
☐ Prepare orchard floor for harvest	30
☐ Apply nitrogen fertilizer in July (complete application, or second half of a split treatment)	27

MANAGEMENT ACTION	Page
FALL (Aug-Oct)	
☐ Monitor webspinning mites	78
☐ Harvest	30
☐ Monitor vertebrates (ground squirrels, birds), control if needed	39,43
☐ Apply potassium (soil) if needed	27
☐ Apply zinc spray at leaf fall if needed	27
☐ Irrigate if needed following harvest	24-27
☐ Apply preemergence herbicides	139
☐ Monitor for peach tree borer	86
☐ Remove weak trees, backhoe, and fumigate tree sites	18,23 131

Figure 8. Seasonal checklist of management actions for an almond orchard and pages in this manual where activity is discussed in more detail.

these records provide valuable historical data for long-term orchard management.

Monitoring Pests

The frequency of monitoring varies with the individual pest species and time of season. Monitoring programs for most species are needed only at certain times of the year. Start monitoring well before populations begin to build and continue through the pest's damaging stages. Sample every orchard for insect and mite pests and their natural enemies at least once a week when pests are likely to be present and more frequently when insect and mite populations can be expected to flare up.

No weekly sampling programs have been developed to assess populations of weeds, diseases, nematodes, and vertebrates. It is important, however, to monitor their presence and watch for changes in their status. Prepare a record once each quarter (fall, winter, spring, and summer) for each orchard, noting the dominant weeds, diseases, vertebrates and any other significant stresses. Correct identification is essential. Sample for nematodes when you suspect they are causing losses or when the site history indicates they may become a problem.

Specific monitoring procedures are described in the discussions of individual pests.

Figure 9. A maximum-minimum thermometer (top) records the highest and lowest temperatures over a 24-hour period. A hygrothermograph (below) records temperature and humidity continuously over several days.

Monitoring Weather

Weather greatly influences the development of both the almond tree and its pests. Good weather information is necessary for timing cultural operations and pest control activities. Many local newspapers and radio stations report daily high and low temperatures, relative humidity, wind conditions, and often evapotranspiration data. The National Weather Service broadcasts local and regional weather information in California over two very high frequency (VHF) FM bands, 162.40 and 162.55 megahertz. Broadcasts include climatological data, frost warnings, agricultural weather forecasts and advisories, river conditions, as well as emergency warnings of hazardous conditions. Taped messages are repeated every 4 to 6 minutes 24 hours a day and revised every 2 to 3 hours.

Weather conditions, however, particularly temperature and rainfall, may vary considerably within a few miles owing to differences in elevation, storm paths, or bodies of water present. Radio and newspaper readings are not recommended for use in calculating degree-days for pest development, for example, because of this local variation. To have the most accurate data for your orchard, set up your own basic weather station. This would include a maximum/minimum thermometer or a thermograph, and a precipitation gauge. A maximum/minimum thermometer records the highest and lowest temperature of the day, whereas a thermograph gives a continuous temperature reading for 24 hours or longer (Figure 9). This information is used to calculate degree-days (see following) and to monitor temperatures on nights of potential frost hazard. A precipitation gauge, which measures the amount of rainfall, is important in setting up an accurate water budget. Place thermometers 5 feet high (1.5 m), and protect them from direct sunlight. A simple canopy shelter is sufficient for thermometers, but thermographs should be enclosed in a weather shelter (Figure 10).

Instruments to measure evaporation (a Class A evaporation pan), humidity (a dry/wet bulb thermometer or a hygrothermograph), wind speed (anemometer), and wind direction (wind vane) are also useful. If a hygrothermograph is used, it should also be enclosed in a weather shelter. Maintain all instruments according to the manufacturer's instructions and calibrate them regularly to assure accuracy. Keep records of all your observations.

Accumulating Degree-Days

The concept of degree-days is important in understanding crop and pest development. Currently, degree-days are used to forecast the development of peach twig borer, San Jose scale, and oriental fruit moth. Research is underway to establish degree-days as a tool for forecasting navel orangeworm and nut development.

Growth rates of the tree and its invertebrate and microbial pests are closely related to temperature. For instance, in a warm year both will develop faster than in a cold year. Therefore, the calendar date alone is not a precise guide for carrying out crop and pest management activities. For more accurate predictions of crop and pest development, time and temperature have to be considered together. A tree or pest's rate of development depends on the amount of heat experienced in the range between its lower and upper developmental thresholds: the minimum and maximum temperatures required for growth. Measuring the amount of heat accumulating over time provides a "physiological time scale," which is biologically more meaningful and more accurate than calendar days. The unit of the physiological time scale is the degree-day (°D)—defined as one degree above the developmental threshold maintained for a full day. You can measure the temperature in Fahrenheit or Celsius; all degree-day figures in this manual are based on the Fahrenheit unit.

Each species of plant and insect has a specific lower and upper developmental threshold. The lower threshold is the temperature below which no development occurs. Above this threshold, the rate of development increases with temperature at a roughly linear rate until it reaches an upper developmental threshold. Above the upper developmental threshold, the growth rate will eventually drop as temperature increases.

To use degree-days for keeping track of crop or pest development, you must know the number of degree-days (amount of heat) required to complete each growth stage. At high temperatures, degree-days accumulate faster, reflecting faster development.

You can roughly estimate degree-days from the daily minimum and maximum temperatures as outlined in the following example:

Add daily minimum and maximum temperatures and divide by 2 to get the average daily temperature.	$\frac{54°F + 74°F}{2} = 64°F$
Subtract the lower threshold temperature (e.g., 50°F for peach twig borer) from the average daily temperature to get the number of degree-days accumulated that day.	$64°F - 50°F = 14°D$

This formula provides an approximate guide but is less accurate if the minimum temperature is below the lower threshold or the maximum temperature is well above the upper threshold. More precise formulas are available for programmable hand calculators and computers.

Figure 10. Protect your instrument for measuring temperature and humidity in a vented weather shelter.

The IPM computer of the University of California is programmed to provide degree-day tables for organisms whose developmental threshold temperatures are known. Several of these tables have been included in the insect and mite chapter under the individual pests. These tables are also available from your farm advisor. To get accurate predictions, use temperature readings taken in your orchard or a nearby, similarly managed orchard with trees of a similar age. Radio or newspaper reports or readings from a more distant orchard are likely to be too different to predict pest development accurately.

Formulas and diagrams of the different methods of computing degree-days are explained in *Degree Days: The Calculation and Use of Heat Units in Pest Management* listed in References.

Control Action Guidelines

A threshold level for an insect or mite pest is reached when economic damage caused by that pest will exceed the cost of control, including the cost of additional applications due to upset of biological control of other pests and incidental expenses.

Control action guidelines indicate when to take management actions to avoid economic crop losses. Use them only in conjunction with careful field monitoring and accurate pest identification.

For most diseases, weeds, nematodes, vertebrates, and some insects and mites, the guidelines are currently less well defined and may be based on combining information about the extent of pest infestation, the potential for effective and economical control, the history of an orchard or a region, the stage of crop development, and the weather.

Management Methods

The best IPM program prevents pest outbreaks and provides long-term, economical control. Preventive aspects of a good program include selecting a deep, well-drained orchard site with light- to medium-textured soil, properly preparing it before planting, choosing cultivars and rootstocks with some resistance to potential problems, removing sources of new infection, choosing the appropriate irrigation and orchard floor management methods, and timing various cultural activities to discourage pest development and keep trees growing vigorously. Pesticides should be used only when careful orchard monitoring indicates they are needed to prevent economic loss, or if crop history indicates preventive sprays are necessary.

The following sections briefly describe common cultural practices associated with almond production. For more information, consult the references listed in the back of this manual.

Site Selection and Preparation

Because healthy, vigorous trees are best able to withstand external stresses, almond trees should be planted where they will be able to develop well both above and below the ground. Many site problems, if they are detected before planting, can be overcome by careful preplant preparations including soil modification, planting on berms or mounds, eradicating perennial weeds, fumigation, and choosing rootstocks or cultivars that offer some resistance to site-related problems.

Almond trees grow on a wide variety of soils but produce best in deep, well-drained soils. Almonds usually do not do well in heavy, poorly drained or stratified soils that restrict root growth. Modify problem soils before planting by deep ripping soils that have a hardpan, or slip plowing or backhoeing stratified soils that have no hardpan. If a clay pan is present, backhoe individual tree sites or slip plow the orchard.

Fine-textured, poorly drained soils provide ideal conditions for the development of Phytophthora crown and root rot, which requires soil saturation or standing water to reproduce and infect host plants. Avoid planting on these soils when possible, especially in areas of higher rainfall (Sacramento Valley). If problem areas must be planted, consider using resistant rootstocks, planting on berms or mounds, and careful leveling of the soil to avoid areas in the orchards where water might accumulate. Sites that have a history of periodic flooding should not be planted as orchards.

If possible, choose an orchard site that does not have a history of cotton, tomatoes, melons and other crops that build up *Verticillium* in the soil. Otherwise, young trees planted in soils where *Verticillium* fungi are present may develop the disease.

Avoid sites that have a history of oak root fungus (Armillaria root rot). This fungus can survive for many years in dead or living roots of fruit or nut trees, oak, aspen, and several fir and willow species. Once a tree is infected, little can be done to save it. If your orchard site has a history of oak root fungus, you will need to thoroughly fumigate the soil before planting (see management guidelines for oak root fungus), and plant your trees on Marianna 2624 rootstock. Even these precautions may not prove successful, however.

Fumigation may also be necessary in orchards where the soil contains root knot, root lesion, ring or dagger nematodes. Sample the selected orchard site to determine if these nematodes are present. This is particularly important if the orchard has sandy or light-textured soils.

Fumigants may be applied as solid, strip or spot treatments. All three methods can be used for nematode control, but only the first method is useful to control oak root fungus. Even then, fumigation for oak root fungus is not always successful, especially in deep, well-drained soils.

Solid treatment. Treat the entire orchard site. This approach is necessary where oak root fungus is a problem and may be necessary for nematode control.

Strip treatment. Treat an area 6 to 8 feet (1.8 to 2.4 m) wide down the proposed tree row. This method may be used where nematodes are a problem. It allows the trees to get off to a good start before nematode populations rebound in the root zone.

Spot treatment. Inject the fumigant at each tree site. This approach is the least expensive because only a limited area is treated. It is sometimes done in conjunction with backhoeing. Spot treatment is recommended when replanting individual trees in an orchard.

For additional information on fumigation, see page 131.

Survey a potential orchard site for the presence of perennial and other troublesome weeds. Perennial weeds are often difficult to control and may require several treatments; they are easier to control before the orchard is planted.

Evaluate the orchard location for potential vertebrate problems (see chapter on vertebrates). Is the surrounding environment ideal habitat for ground squirrels, gophers, birds, rabbits, or deer? If so, can the habitat be altered to make it less favorable for these pests? Abandoned or poorly managed orchards, rangeland, wooded areas or stream banks located close to an orchard may harbor vertebrate pests. Preventive action taken at planting, such as fencing the orchard or protecting individual tree trunks, may be necessary to prevent tree damage in orchards located near good vertebrate habitat.

Once a site is selected, prepare the land in the fall before spring planting. Fall is the best time for land preparation because land leveling, subsoiling, backhoeing or slip plowing need to be done when the soil is dry. It also allows adequate time before planting for the ground to resettle. The land should be prepared early enough in the fall to allow time for fumigation and other activities. Any low areas with poor drainage need to be corrected. Irrigation systems should be designed at this time, although some systems can be installed after the trees are planted. Not only is proper land preparation important for optimum root growth by the trees, it is also important for optimum performance of fumigants and herbicides used in the orchard and to minimize potential phytotoxic effects of the chemicals on the trees.

Cultivar and Rootstock Selection

Potential for pest damage is greatly influenced by choice of rootstocks and scion cultivars. At present, five rootstocks are available for use with almond trees. Each one has advantages and disadvantages associated with its use (Figure 11). Before planting an orchard, the soil properties and the history of the site should be thoroughly investigated so that the selection of a rootstock takes into consideration potential pest problems along with horticultural concerns, such as scion compatibility.

The five rootstocks used in almonds are almond, Lovell peach, Nemaguard peach, Marianna 2624 (plum) and almond-peach hybrid. Myrobalan plum and apricot rootstocks were used for almonds in the past, but both were unsatisfactory. Lovell peach and Nemaguard peach are the most commonly used rootstocks of the five available. Nemaguard peach is used primarily where root knot nematode is a problem in sandy soils and in areas of low rainfall. Lovell peach, on the other hand, is used mainly in well-drained loam to clay loam soil, in areas north of Sacramento, and in areas where root knot nematodes have not been serious. Marianna 2624 is tolerant of wet soil conditions and resistant (but not immune) to oak root fungus and root knot nematodes. It is compatible with some cultivars such as Mission, Padre, Ne Plus Ultra, Peerless, Thompson, Carmel and Harvey, but in-

SUSCEPTIBILITY[1] TO—	Almond	Lovell peach	Nemaguard	Marianna 2624	Almond-peach hybrid[3]
Phytophthora crown and root rot	ES	MS	S	SR	VS
oak root fungus	S	S	S	SR	ND
crown gall	VS	S	S	SR	ND
Verticillium wilt	VS	VS	VS	SR	VS
root knot nematode	VS	S	R	R	SR
other nematodes[2]	S	S	S	S	S
nutritional disturbances	4	5	5	ND	6
peachtree borer	S	ES	ES	ND	ND
pocket gophers and mice	VS	SR	SR	ND	ND

1. ES – extremely susceptible, VS – very susceptible, S – susceptible, MS – moderately susceptible, SR – somewhat resistant, R – resistant, ND – no data.
2. Other nematode pests of almonds include root lesion nematode, *Pratylenchus vulnus*, ring nematode, *Macroposthonia xenoplax*, and dagger nematode, *Xiphinema americanum*.
3. Variable in response depending on the particular hybrid used. Not well documented.
4. More resistant to excess boron and lime-induced chlorosis than other rootstocks listed.
5. More sensitive to calcareous soils and high boron or chloride than trees on almond rootstock.
6. More tolerant of lime, sodium and chloride than Lovell peach but highly sensitive to excess boron.

Figure 11. Susceptibility of major rootstocks to pathogens, nematodes, and nutritional disturbances.

compatible with Nonpareil, Solano, Milow, and some older cultivars such as Kapareil and Drake. The compatibility of Marianna 2624 rootstock with some of the newer cultivars is not clear at this time; check with your nurseryman. Finally, trees on Marianna 2624 are smaller than those on Lovell peach and may be shorter lived. Because of these limitations, Marianna 2624 is best used in spot situations where other rootstocks would not survive. If an entire orchard is to be planted on Marianna 2624, space the trees closer together than you would for other rootstocks.

Trees planted on almond or on almond-peach hybrid rootstocks also have limited ranges. In addition, the range for almond-peach hybrid rootstocks may vary depending on the kind of hybrid.

Trees on almond rootstock should be planted in deep, well-drained soils where root knot nematode is not a problem. Because of the deep-rooting characteristic of this rootstock and its susceptibility to Phytophthora crown rot, trees on almond rootstock must be planted on land where the water table is deep and drainage is good. Almond-peach hybrid rootstocks produce large, vigorous, deep-rooted trees. These trees are somewhat more tolerant to drought conditions during late summer and fall and of calcareous soils than trees on Lovell peach rootstock. However, trees on this rootstock may become so vigorous when grown in deep, fertile soil that framework limbs tend to split during the development period, and overall size could become a problem. In addition, trees on this rootstock should not be planted where crown rot conditions exist. For these reasons, trees on almond-peach hybrid rootstock may be best suited to well-drained, shallow, or sandy soils where their extra vigor may counteract soil limitations.

Many almond scion cultivars have been introduced during the past 25 years, and additional cultivars will be forthcoming in the future. The major scion cultivars currently growing in California include Nonpareil, Mission, Merced, Ne Plus Ultra, Thompson, Peerless, Price, and Carmel. When planning an orchard, consider the disease and insect susceptibilities of the different scion cultivars (Figure 12) along with horticultural considerations such as bloom time, time and ease of harvest, nut quality, and marketability. If the orchard is located in an area where a particular disease is known to be damaging, select cultivars, when available, that have some resistance to that disease. Frequently, growth characteristics of the tree or properties of the nut can influence the potential for pest damage. Navel orangeworm, peach twig borer and oriental fruit moth, for example, are most damaging to cultivars with nuts that have poor seals or soft shells or to cultivars that have an extended hull split period. Finally, the cultivars and rootstocks chosen for an orchard (as well as the soil they are planted in) may also influence the selection of herbicides that are used in the orchard. For example, Mission trees on Marianna 2624 rootstock are more sensitive to simazine damage than other cultivar/rootstock combinations, particularly if they are growing in sandy soils that are low in organic matter.

Planting and Managing a New Orchard

When you receive young trees from the nursery, do not allow the roots to dry out or to be exposed to freezing temperatures. If you cannot plant the trees immediately, heel them in damp soil or damp sawdust, but do not cover the tree trunks above the depth at which they grew in the nursery. When heeling-in young trees, observe the following precautions: First, don't let the heeling-in area become waterlogged or the trees may develop bundle rot (see Phytophthora crown and root rot). Keep the roots moist but not wet, and avoid large air pockets around the roots. Second, make sure the soil in the heeling-in area is free of disease organisms. If you fumigated your orchard site, heel the trees in soil that has been fumigated but is free of herbicides. Take the same precautions with the soil placed around the roots when you plant the trees. As soon as soil conditions permit, plant the trees.

Tree hole sites are generally marked and dug with an auger. In sandy to loam soils it is often nearly as fast, and in some cases better, to dig the holes with a shovel. Be sure not to dig the holes any deeper than is necessary to cover the roots. Trees planted in holes that are too deep may settle in the soil and develop crown rot. Because trees will settle deeper into the soil after irrigation, plant them high, with the uppermost root just under the soil surface. After settling, the trees should stand no deeper than they did in the nursery.

If the soil is too moist, the auger may cause the sides of the hole to seal. Or, if the hole was dug a day or so before planting, the sides and bottom of the hole may crust. If either situation has occurred, be sure to break down the sides of the hole before planting to prevent confinement of the tree roots or poor water penetration. Trees should be watered the same day they are planted, unless the soil has very good moisture. Dry soil, air pockets, and poor soil contact resulting from insufficient water can kill newly planted trees. Even a rain immediately after planting, unless it's quite heavy, is not sufficient to settle the soil around the root system. However, trees can also be lost from excess moisture caused by tanking already moist soils or excessive rain occurring after tanking.

At planting time, trim off any roots that are smashed, broken, or too long to fit in a reasonable size planting hole without being bent. If crown gall is a problem, dip the remaining root system in a solution of Galltrol, a commercially available biological agent that prevents infection by crown gall.

SUSCEPTIBILITY[1] TO—	Non-pareil	Merced	Mission	Peerless	Ne Plus Ultra	Butte	Carmel	Fritz	Price	Ruby	Thomp-son	Padre	Solano	Sonora
almond leaf scorch	VS	SR	SR	VS	S	R	R	R	R	R	R	SR	SR	SR
bacterial blast	ND	ND	ND	ND	VS	ND	ND	ND	ND	ND	ND	ND	ND	ND
brown rot	SR	S	S	S	VS	VS	VS	S	SR	VS	S	S	SR	S
ceratocystis	VS	S	ES	S	S	S	S	S	S	S	VS	ND	ND	ND
band canker	S	ND	S	ND	S	ND	S	ND	ND	ND	ND	ND	ND	ND
foamy canker	S	ND	S	S	S	VS	VS	ND	S	ND	ND	ND	ND	ND
hull rot	VS	S	SR	SR	S	S	S	S	S	S	S	SR	S	ND
leaf blight	S	VS	S	VS	VS	S	ND	VS	ND	ND	S	ND	ND	ND
scab	S	VS	S	S	VS	S	VS	S	VS	S	S	ND	ND	ND
shot hole	VS	ES	S	VS	ES	S	S	S	S	S	ES	S	ND	ND
Verticillium wilt	S	SR	SR	S	VS	S	VS	S	S	S	S	ND	ND	ND
yellow bud mosaic	SR	R	VS	R	SR	R	R	R	R	R	R	ND	ND	ND
mites	VS	VS	S	S	VS	VS	VS	VS	S	S	S	SR	S	ND
navel orangeworm and peach twig borer	S	VS	R	SR	S	SR	SR	S	SR	SR	VS	R	SR	S

1. Cultivars are rated as ES (Extremely Susceptible), VS (Very Susceptible), S (Susceptible), SR (Somewhat Resistant), and R (Resistant). ND means no data is available.

Figure 12. Susceptibility of major scion cultivars to pathogens and other pests.

After planting the trees, head them back to about 30 inches (76 cm) and place a tree protector or a quart milk carton around the tree. If you don't plan to use a tree protector or milk carton, plant the trees so that the bud union is on the southwest side to help reduce sunburn on the stock. You can also protect the tree trunks from sunburn, especially if the trees are planted late, by painting the trunks. Where prevailing winds are a problem, plant trees at a slight angle into the wind.

Planting on Berms or Mounds. On heavier soils, trees survive better when planted on berms or mounds. Planting on berms or mounds will improve drainage around the crown of the tree and help reduce the risk of crown rot. Berms also reduce chances of tree damage due to certain herbicides and reduce weeds around tree trunks because berms remain dry during irrigation.

Use a berm on leveled ground with a slight slope and prepare it before planting. The berm should be about 2 to 3 feet wide (60 to 90 cm) and 6 to 8 inches (15 to 20 cm) higher after settling than the surrounding soil, unless irrigation practices dictate an even higher berm. On uneven ground, where berms are not feasible, plant the trees on mounds. The mound should be about 3 feet (90 cm) square and about 6 inches (15 cm) high.

Irrigating a New Orchard. If the soil has good moisture at planting time, work the soil around the roots by hand

and do not irrigate until after shoot growth has started. If the soil is dry at planting time, apply water to settle the soil around the roots. Be sure the planted trees do not settle too deeply in the soil. If you use water to settle the soil around a tree, do not firm the soil further by stepping on it. This compacts the soil and may interfere with root development and water penetration around the base of the tree. Do not irrigate again until after the trees have grown about 4 to 6 inches (10 to 15 cm).

Be sure to supply adequate water to trees during the first year. Use frequent, light irrigations, particularly on shallow or sandy soils. Almond trees planted in a deep soil that has been wetted to field capacity by rains or irrigation may need less frequent irrigations the first year, provided the soil is kept absolutely weed free to eliminate competition for water. Under such circumstances, trees are able to get part of the water they need as the root system expands into new moist soil.

Lack of readily available moisture can seriously reduce the growth of newly planted trees. If trees remain dry for even a short time during the summer, they may be weakened, sunburned, and/or attacked by wood borers. Dry trees may not begin growing again even if subsequently irrigated.

During the remaining nonbearing years, adequate soil moisture is the most important consideration in obtaining good tree growth. When the soil has not been wetted by winter rains, you may want to give a partial irrigation to the orchard before the start of the growing season. In areas of normally high rainfall, be careful that an early irrigation does not induce root and crown rot. Control weeds to reduce competition for both soil moisture and nutrients.

If you irrigate with surface or sprinkler systems, plan your last irrigation for early August in the Sacramento Valley or late August in the northern San Joaquin Valley. If you irrigate first- and second-year trees much later than late August in the Sacramento Valley or mid-September in the northern San Joaquin Valley, the trees may be killed by Phytophthora crown and root rot. Substantial early fall rains can cause similar damage. (Fall irrigations in the southern San Joaquin Valley, where rainfall is considerably lower than the northern areas, do not generally cause crown rot problems.) Crown and root rot damage becomes evident in the late spring following infection when the trees, which had started growth, suddenly stop growing, the leaves turn yellow and then brown, and the trees die. It is safer to keep a young orchard weed-free in the fall and to allow the trees to use moisture remaining in the soil from earlier irrigation than to risk a *Phytophthora* infection by irrigating too late.

Pruning (Training) a New Orchard. Prune the tops of newly planted small trees (3/16 to 3/8 inch or 5 to 10 mm trunk diameter) by topping the trunk with one cut about 32 inches (81 cm) above ground. The primary scaffolds, which develop within 6 to 10 inches (15 to 25 cm) of this cut, will form the tree crotch. Because mechanical shakers are commonly used for harvest, you must provide sufficient unbranched trunk space between the soil and the lowest scaffold for shakers to attach to. Larger trees (1/2 inch or 12 mm in diameter and larger) with lateral branches along the trunk should be topped at about 32 to 36 inches (81 to 91 cm). Lateral branches can be removed, or if they are in desirable locations, they can be headed back, leaving 1 to 2 inches (2.5 to 5 cm) of growth with one or two lateral buds.

After the first growing season, the primary scaffold limbs of the tree are selected by pruning. Because almond trees normally branch readily, the primary scaffold limbs can be left unheaded or tipped at 36 to 42 inches (91 to 107 cm) from the crotch.

The secondary structure is established following the second growing season. Secondary limbs are usually not headed unless the tree needs balancing or additional branching is desired.

Fertilizing a New Orchard. During the first growing season, trees frequently benefit from light applications of nitrogen fertilizer. Apply 1 to 2 ounces (30 to 60 g) of actual nitrogen per tree after growth is started in either a single (late May) or split (May and early July) application. You can either place the fertilizer in the irrigation furrows, or if using sprinkler or flood irrigation, broadcast it around the tree. However, always keep it at least 18 inches (46 cm) or father from the trunk. Do not put fertilizer in the holes at planting time.

Intercropping. In young orchards, an annual crop is sometimes grown between the tree rows for 1 to 2 years to obtain additional income. Crops must be chosen with great care or the income may be negligible and the almond trees may be damaged. The welfare of the trees must always come first.

If the orchard is intercropped, do not use crops that require the soil to be dried out during the summer so that they can be harvested. Examples of such crops are grains and oat hay. Drying can seriously stunt almond trees. In addition, if the drying crop catches fire, the orchard trees can be seriously damaged or destroyed.

Many broadleaf plants, including cotton, lettuce and alfalfa, are good hosts for root knot nematodes. Infestations of these broadleaf crops in sandy soils can produce high nematode populations so rapidly that they will reduce growth of the almond tree in spite of the good "field resistance" available from Nemaguard rootstock. Mites also can be serious in intercropped orchards, especially if beans are planted as the intercrop.

In addition, Verticillium wilt can damage young almond trees; it nearly always builds up in the soil when tomatoes, cotton, or related crops are grown. These crops, therefore, may not be good intercrops.

Any herbicide or other pesticide you use on the intercrop may damage the almond trees. Moreover, because of the presence of the intercrop and because of residue problems on it, you may not be able to use the most suitable pest control material on the almond trees. Finally, care must be taken that the equipment used for the intercrop does not damage the trees.

Pest Management in Nonbearing Orchards. Once the orchard is planted, protect young trees from sunburn and wood borer damage by wrapping the trees with a protective tree wrap or painting the trunk of the trees with white interior latex paint. Because the crown of the tree is extremely susceptible to wood borer damage, dig down slightly below the soil line and paint exposed surfaces. After painting, place the soil back around the trees making sure the painted surface extends below the soil line.

Monitor your trees in spring for the presence of disease and pest problems. Although it is usually not necessary to control diseases such as brown rot or shot hole the first year, the trees may need to be protected from peach twig borer. If the trees did not receive a dormant treatment in the nursery, treat the trees in March just after they leaf out. If the trees did receive a dormant application, monitor them in mid-April for peach twig borer or oriental fruit moth shoot strikes. Both insects damage young trees by destroying terminals and causing buds below the terminal to grow. Excess lateral growth often makes pruning and shaping of the tree extremely difficult. If shoot strikes are common, treat the trees according to the guidelines given for spring sprays for peach twig borer. Use your monitoring results to develop disease and pest control strategies for the following year.

If you are planning to use nontillage or only disc the orchards in one direction and use strip weed control down the row, herbicide applications are generally safe after the first growing season. When using a contact (postemergence) herbicide, make sure it doesn't come into contact with the tender bark on young trees. Preemergence herbicides should be applied in the fall following the first growing season before winter weeds and grasses begin to grow and rain, or sprinkler irrigation, is available to move the herbicides into the ground.

One-year-old (first leaf) trees should receive a dormant treatment of an organophosphate insecticide and oil to control peach twig borer, San Jose scale, and European red mite and brown mite eggs. The same dormant treatment should be applied to 2- and 3-year-old trees.

Handle second and third leaf trees the same as newly planted trees, except that in prolonged wet springs, a fungicide for control of shot hole may be needed if symptoms of the disease occurred in the orchard the previous year. If moderate shoot damage was experienced the previous season, continue to monitor for peach twig borers and oriental fruit moth in mid-April.

In certain areas of the state, American plum borer is a serious problem on 2- and 3-year-old trees. If American plum borer becomes a problem, follow the guidelines for this pest in the insect and mite chapter of this manual.

Replanting in an Established Orchard

When replanting in an established orchard, use a backhoe to remove the old tree stump and as many of the large roots as possible. The backhoe will also cut the roots of surrounding trees that would compete with the replant. Carry out this operation in fall before planting to allow time for soil to settle. Then refill the hole, fumigate, settle the soil, and make the planting hole in the usual manner at planting time.

Whatever the reason for replanting trees in an established orchard, it is frequently desirable to spot fumigate the soil before planting the new tree. If correctly done, spot fumigation gives the new tree an area relatively free of nematodes and disease organisms in which to make a start. Fumigation also kills roots of older trees that have grown into the new planting location. By reducing competition for nutrients and water, the young replant has a much better chance of doing well. Fumigation can be hazardous, so follow the precautions on the label.

If root knot nematodes are a severe problem in the orchard, use a rootstock that is resistant to this nematode, even if you fumigate the soil. Similarly, use a rootstock that is tolerant of oak root fungus (Armillaria root rot) if this disease is the reason for replanting. (See *Cultivar and Rootstock Selection,* page 19.)

Replants will not do well if they are heavily shaded by surrounding trees. If shading is a problem, cut back the tops of surrounding trees.

While it is important to supply water and nutrients to any newly planted trees, it is essential in the case of a replant because of competition from the older surrounding trees. Give replants extra irrigations, at least during the first 2 years, but do not allow the soil to become too wet or the trees may develop crown and root rot. Also apply small amounts of a nitrogen fertilizer once or twice during the summer to help overcome competition. Apply nitrogen fertilizers at least 18 inches (46 cm) from young trees. Do not allow weeds to grow around a replant; they can become severe competitors for water and nutrients.

Frost Protection

The dormant almond tree is quite tolerant of temperatures below freezing. Because almonds bloom earlier than most other deciduous fruit trees, however, the blossoms and young nuts have a greater likelihood of being damaged by frost.

Sprinkling under the tree during bloom will protect almond blossoms from frost damage, if temperatures do not drop more than a few degrees below freezing, but it may also increase disease problems due to continual wetting of blossoms and fruit in the lower parts of the tree. Also, if frost occurs several nights in a row and if soil drainage in the orchard is restricted, root drowning and rotting could occur.

The presence or absence of a cover crop can influence the frost hazard potential of an orchard. Bare, dark, moist, firm soil stores more heat during the day for release at night than soil with a cover crop or soil that has been cultivated and left loose. A closely mowed cover crop, however, is almost as warm as bare, moist, dark soil. For more detailed information on frost protection in almond orchards, see *Almond Orchard Management* listed under General References.

Water Management

Adequate quantities of water are essential for growth and crop production in almonds. Allowing an almond tree to become stressed for water early in the season will slow shoot growth. This is especially serious in young trees where maximum shoot growth is needed to hasten orchard development. Early season moisture stress will also lower yields by reducing nut enlargement. Severe water stress during the summer may cause leaf drop and sticktight nuts. Too much water, on the other hand, damages roots by depriving them of oxygen and creates conditions that favor infection by soilborne disease organisms. Phytophthora crown and root rot in particular, one of the most destructive diseases of almonds, is favored by saturated soils.

Irrigation Methods. Several methods can be used to irrigate almond orchards, including surface (flood, furrow), sprinkler, and localized (drip, low-volume sprinklers) irrigation. Presently flood and sprinkler irrigation are the most common, although each system can supply water adequately if managed properly. Major considerations in choosing a system are soil texture, slope of the land, and the cost, source, and availability of water. Installation and operation costs are also important, as well as frost hazards in the orchard.

Surface Irrigation. Irrigation water can be most rapidly applied on level or leveled-to-grade soil by using furrow or flood irrigation if a good head of water is available. Rapid application of a large amount of water can be helpful in extremely hot weather if the need arises for immediate irrigation. Also, because gravity is used to distribute water across the orchard, energy inputs with surface methods are relatively small. However, because it takes some time for water to reach the far end of the field, usually more water will infiltrate the soil at the head end of the orchard. Also, because soil infiltration rates can vary greatly throughout the orchard, uniformity of infiltrated water can be low with surface irrigation. Finally, the temporary levees used for surface irrigation require considerable labor and may make it difficult to smooth the soil for mechanical harvesting. If the soil-smoothing operations are not well timed, clods left on the surface may interfere with harvesting. Under nontillage, however, soil smoothing is not a problem in flood-irrigated orchards where berms are present.

Use of surface irrigation may affect certain pest problems. For example, if water comes from canals, it may be contaminated with weed seeds and pathogens, which can spread through the orchard as the water moves across the soil surface. On the other hand, flood and furrow irrigation may discourage ground squirrels and pocket gophers from digging burrows in the orchard.

Sprinkler Irrigation. Properly designed sprinkler systems apply water fairly evenly over the soil surface and the orchard floor requires minimal soil-smoothing operations to produce a smooth surface for mechanical harvesting. In addition, sprinklers require less labor than do most other methods of irrigation. (Considerable energy is required, however, to distribute the applied water.) Other advantages of a well-designed sprinkler irrigation system that is properly managed include frost protection and distribution of fertilizers and herbicides through the system. Because sprinklers usually apply water at a rate lower than the soil infiltration rate, uniformity of water infiltration is primarily determined by the system design. Thus, with sprinkler irrigation, water infiltration is generally more uniform throughout the orchard than it is with surface irrigation methods. Sprinkler systems with low rates of application can be used on land where runoff or slow penetration would be a problem if surface irrigation methods were used.

The use of a sprinkler irrigation system, however, can increase disease problems if water is applied so that leaves or nuts are wetted.

Localized Irrigation. Localized irrigation, which includes drip (trickle) and low-volume (micro, mini, mister, fogger) sprinklers, is another method used for almonds. With localized irrigation, soil moisture used by the tree is replaced by applying water to a relatively small percentage of the orchard floor on a frequent (often daily) basis. Trees irrigated by this method have a reduced root zone compared to trees irrigated by other methods. Also, water volume in the soil surrounding the roots of localized irrigated trees is lower than with other methods. As a result, the soil water reservoir must be refilled frequently.

Advantages of this system are potential water savings in the early years of the orchard and usually minor interference with most cultural operations. In mature orchards, water savings under localized irrigation may occur because application is more efficient with other methods. Localized irrigation usually works well on soils with a low infiltration rate and is well suited for automatic operation with time clocks.

The disadvantages of localized irrigation are high costs, maintenance, emitter clogging, lack of reserve moisture in the soil, and some possible interference with harvest. Clogging, a major difficulty, can usually be reduced by filtering or by chemical injection, depending on whether it results from physical particulates, chemical precipitates, algae or slimes.

Drip irrigation must be engineered to meet the maximum expected water use demand, or trees can become stressed and production may be reduced more rapidly than with other irrigation methods. Also, drip irrigation will hinder weed control because it keeps the area around the emitters moist, facilitating rapid breakdown of preemergence herbicides.

Recent research indicates that low-volume sprinklers, which wet a larger surface area than drip emitters, but on a less frequent basis, allow the soil surface to dry and thus increase herbicide effectiveness.

Scheduling Irrigations. Water budgeting is the primary method used to determine when to irrigate and how much water to apply. To use this method, you estimate how much water the trees are using and then irrigate on a timely basis to meet their water needs. You also need to know soil water-holding capacity, root zone depth, and irrigation efficiency. Another principal method of irrigation scheduling involves making soil moisture measurements and attempting to relate them to the well-being of the trees.

Water budgets can be used with all irrigation methods but are easiest to use with localized irrigation because soil water-holding capacity and root zone depth are of minor importance in this case. Also, irrigation efficiency for localized irrigation is simpler to compute than with other irrigation methods. Water budgets are not recommended for use in scheduling irrigations on soils with high water tables or in areas where reliable evapotranspiration (ET) data are not readily available.

The first step in creating a water budget is to determine how much water the soil in the tree's root zone can hold—that is, the size of the soil water reservoir. Begin by finding out how much water is available when the soil is at *field capacity:* the water content of the soil 2 to 3 days after a heavy rain or irrigation (Figure 13). About half of the soil volume consist of solid particles. The rest of the soil volume is pore spaces, holding varying proportions of air and water. The amount of water that a soil can hold

Soil is about half solid material by volume (large circle). The rest of the soil volume consists of pore spaces between soil particles. Pore spaces hold varying proportions of air and water (small circle).

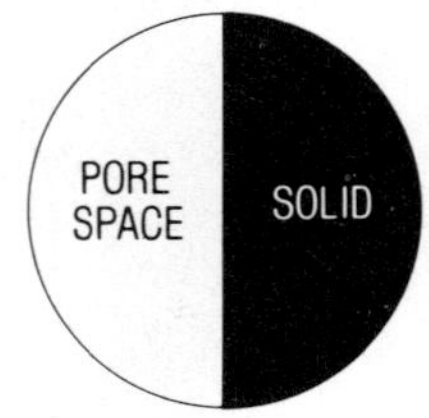

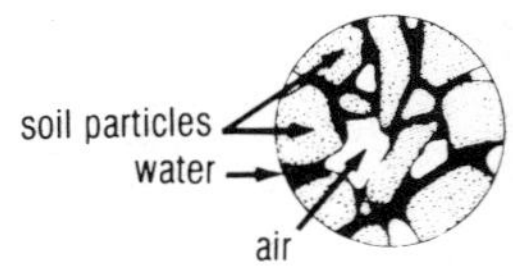

When the soil is **saturated** after irrigation or rain, pore spaces are filled with water.

When soil has drained following irrigation, it is at **field capacity.** In most soils, about half of the pore space is filled with water. About half of this water is **available** to plants; the rest is unavailable because too much suction is needed to remove it from pore spaces. The proportion of soil water that is available is higher in clays and lower in sandy soils.

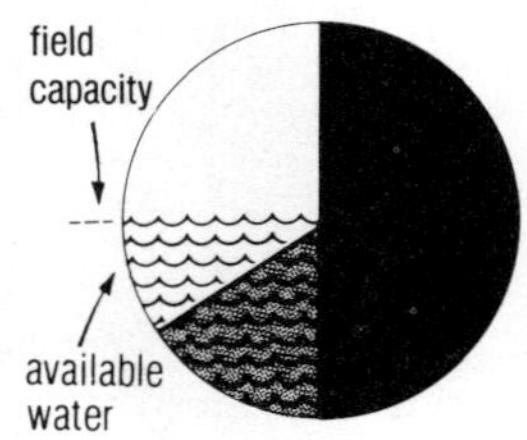

The **allowable depletion** is the proportion of the available water the crop can use before irrigation is needed.

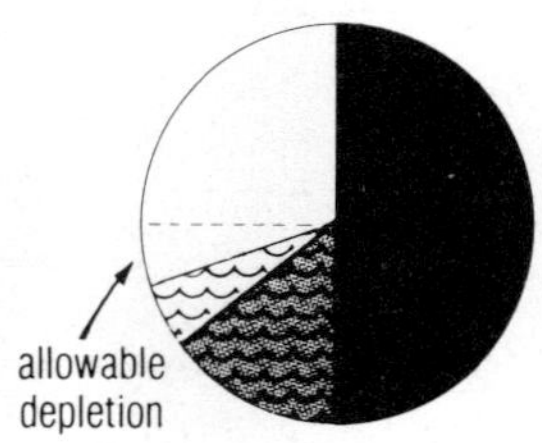

At the **wilting point,** all available water is gone. Plants die unless water is added.

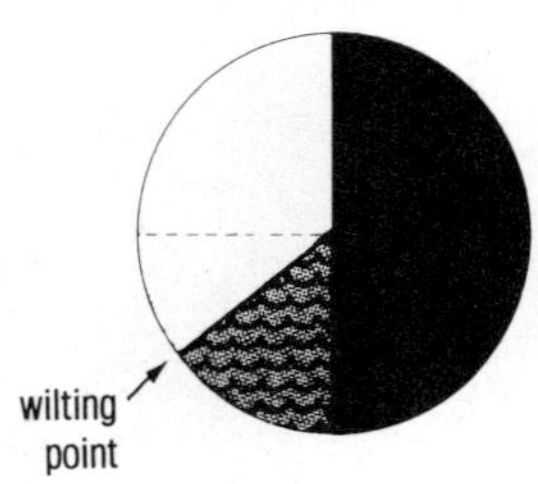

Figure 13. The soil reservoir.

depends, in large part, on the textural characteristics of the soil. Not all of this soil water can be extracted by plant roots, however. The lower limit of water extraction is termed the *permanent wilting point*. The difference between field capacity and the permanent wilting point is the *available water* content. Like field capacity, the available water content of a soil depends largely on soil texture. For example, the amount of available water is greater in clay soils than sands, which have lower total pore space. Table 1 shows the approximate water-holding capacity of various soils. You will need to determine the texture of the soil throughout the root zone of the tree and the total root zone depth. Use a soil auger to examine the soil profile. One way to determine soil texture is to use the feel method described in Table 2. When available, soil maps prepared by the U.S. Soil Conservation Service can be used to obtain information on soil texture. To determine the rooting depth, visually examine the soil core for the presence of tree roots.

Table 1. Available Water for Various Soil Types.

	AVAILABLE MOISTURE	
Type of soil	**Range (in/ft)**	**Average (in/ft)**
Very coarse to coarse-textured sand	0.5 to 1.00	0.75
Moderately coarse-textured sandy loams and fine-sandy loams	1.00 to 1.50	1.25
Medium-texture—very fine sandy loams to silty clay loams	1.25 to 1.75	1.50
Fine and very fine texture—silty clay to clay	1.50 to 2.50	2.00
Peats and mucks	2.00 to 3.00	2.50

Table 2. Guide for Determining Soil Texture Using the Feel Method.

Type of soil	Soil characteristics at field capacity
Coarse (loamy sand)	Leaves wet outline on hand when squeezed.
Sandy (sandy loam)	Appears very dark; leaves wet outline on hand; makes a short ribbon.
Medium (loam)	Appears very dark; leaves wet outline on hand; will ribbon out about 1 inch.
Fine (clay loam)	Appears very dark; leaves slight moisture on hand when squeezed; will ribbon out about 2 inches (5 cm.)

Don't wait until all of the available water in the tree's root zone has been extracted before irrigating because water stess begins before complete depletion. The proportion of available water that can be used before irrigation is needed (*allowable depletion*) depends on the plant species, its growth stage, soil texture, and atmospheric conditions. Although no single recommendation can be made for all orchard situations, generally when 50% of the available moisture has been used, it is time to irrigate.

Once you determine available water content and allowable depletion level of your soil, the total water-holding capacity can easily be calculated by estimating the rooting depth of the tree. Figure 14 shows an example for a profile containing three soil types. Total water-holding capacity is calculated by determining available water in each layer of the root zone. Multiply the total available water-holding capacity by the allowable depletion level to determine the amount of water use permitted between irrigations. When soil moisture loss reaches this predetermined depletion level, you must irrigate to refill the soil water reservoir.

You can estimate water loss by using evapotranpiration (ET) data. ET is defined as the amount of water lost through transpiration by the plant and by evaporation from the soil surface. ET tables based on long-term average conditions are available for mature almond orchards with or without a ground cover. An example is given in Table 3. ET rates 10 to 25% above or below average may occur for short periods in certain years, particularly during spring and fall. Historical data for a normal year provides a better approximation for midsummer when the weather does not vary much from year to year. Current ET data for different crops are often provided in local newspapers, on the radio, or by irrigation districts. By adding up weekly (or daily, if available) ET figures until they approach the calculated allowable depletion, you can estimate when the next irrigation is needed (Figure 15).

To calculate the actual amount of water that needs to be applied, you must determine what portion of the applied water is actually available for use by the trees. Certain unavoidable losses occur during irrigation, mostly from deep percolation and runoff. Surface irrigation is normally 60 to 80% efficient, sprinkler around 75% efficient, and localized irrigation about 90% efficient. Rain is estimated to be 70% efficient. Private irrigation consultants can be hired to assess your irrigation efficiency, or you can learn how to estimate the effectiveness by contacting the U.S. Soil Conservation Service or local farm advisor. Multiply the net amount of irrigation water needed at each irrigation by the calculated amount for inefficiency.

At the end of the week following an irrigation, record the ET amount for the days following irrigation from the newspaper. In midsummer when the need to irrigate may be weekly or biweekly, use long-term weekly

Table 3. Estimated Monthly Evapotranspiration from Almond Orchards in the Sacramento and San Joaquin Valleys, Based on Long-term Weather Data.

	ESTIMATED MONTHLY EVAPOTRANSPIRATION (INCHES)												
Location	Jan	Feb	Mar	Apr	May	Jun	Jul	Aug	Sep	Oct	Nov	Dec	Total
Sacramento Valley													
With ground cover[1]	0.9	1.0	3.3	4.9	7.0	8.7	9.5	8.0	5.9	2.1	1.2	0.8	53.3
Clean cultivated		0.6	2.1	3.4	5.2	6.9	7.6	6.5	4.4	1.4			38.1
San Joaquin Valley													
With ground cover[1]	0.8	1.4	3.2	5.1	7.5	9.0	9.4	7.9	5.6	3.5	1.2	0.6	55.2
Clean cultivated			1.9	3.4	5.4	6.9	7.5	6.3	4.4	2.5			38.3

1. Either a green, actively growing cover crop or weeds. If the ground cover dies or is senescent, consider orchard to be clean cultivated with respect to water use.

average ET information (divide monthly estimates by 4 to get weekly averages) to project the next irrigation date. When the next published weekly ET data becomes available, use it to confirm your projected irrigation date. If you have already irrigated, based on a projected date, and you underestimated the depletion level, make the appropriate adjustment when planning the next irrigation. For example, if you discover when the weekly ET rate is published that you miscalculated the amount of water extraction from long-term data and only replaced 90% of the soil water reservoir, when you start you next water budget, begin with a 10% deficit.

The amount of water you can apply at one time, however, often depends on the intake rate of the soil. For minimum disease problems and maximum orchard life, the irrigation should be absorbed in a maximum of 48 hours. Unfortunately, some California soils do not absorb water at a rate that will resupply the entire root zone within this time restriction. If the soil cannot absorb the needed amount you have calculated within a 48-hour period, you will need to adjust irrigation intervals accordingly.

Soil moisture measurements, the second major approach to irrigation scheduling, can be used effectively to confirm your calculations and verify that the soil profile has been filled. Check the soil moisture after an irrigation and periodically between irrigations with soil tubes, soil augers, gypsum blocks, or tensiometers. Tensiometers, which are installed at different soil depths, tell how tightly water is held in the soil. Quantitative evaluation of tensiometer readings is difficult because they depend on soil type, depth of instrument placement, and irrigation method. Readings can be used qualitatively, however, to check if the water has reached a certain depth after an irrigation (when suction decreases) and how dry the soil gets between irrigations (when suction increases). Tensiometers are particularly useful for localized (drip) irriations because they can indicate whether or not the system is maintaining adequate moisture in the root zone.

Tensiometers work best in sandy soils. Clay and clay loam often break the suction in the tensiometer, so frequent maintenance is necessary. In these heavier soils, gypsum blocks require less maintenance and may be more reliable.

Fertilization

Fertilization is usually necessary to resupply the trees with adequate amounts of all essential nutrients. Trees weakened by nutrient deficiencies produce less and are more susceptible to pest damage than healthy trees. Most California orchard soils contain sufficient amounts of the essential nutrients except nitrogen, which has to be replenished on a yearly basis. If you maintain a ground

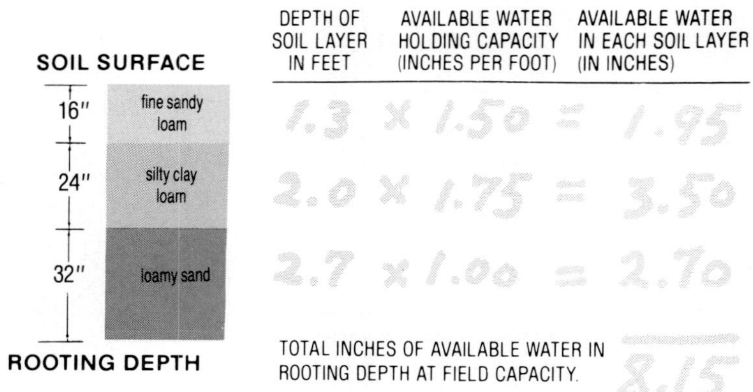

Figure 14. To calculate how much available water the soil can hold at field capacity, first examine the soil profile for layers of different soil textures in the rooting depth of the tree. For each soil layer, multiply the depth of the layer in feet by the available water-holding capacity of that soil texture (see Table 1). Add results from all layers to get the total available water-holding capacity for the soil profile.

ORCHARD/CROP San Joaquin Valley bare soil / Almonds (mature)

SOIL Fine sandy loam, silty clay loam and loamy sand

AVAILABLE WATER 1.00 to 1.75 in./ft. (from Table 1)

ROOTING DEPTH 6 ft.

TOTAL AVAILABLE STORED WATER ≈ 8.0 in. (from Figure 14)

ALLOWABLE DEPLETION 50%

ALLOWABLE LEVEL OF WATER DEPLETION 4.0 in.

IRRIGATING EFFICIENCY 75% (sprinkler)

GROSS IRRIGATION REQUIREMENT 4.0 in ÷ 0.75 = 5.3 in.

DATE	ET (inches)	CUMULATIVE ET between irrigations	REMARKS
Mar. 1-7	0.48	0.48	Assume soil at field capacity from spring
8-14	0.52	1.00	
15-21	0.56	1.21	Rained 0.5 in. Mar. 17; 0.35 in. (70%) subtracted from cum. ET.
22-28	0.65	1.86	
Mar. 29 – Apr. 4	0.79	2.65	
Apr. 5-11	0.85	3.50	Cum. ET approaching allowable depletion; irrigate soon
12-18	0.92	0.42	Irrigated Apr. 16; start new cum ET
19-25	1.25	1.67	
Apr. 26- May 2	1.35	3.02	
May 3-9	1.40	0.42	Irrigated May 8
10-16	1.42	1.84	
17-23	1.67	3.51	Irrigate soon
24-30	1.90	1.41	Irrigated May 25

Figure 15. Sample form for scheduling irrigation using evapotranspiration (ET) data. Start accumulating ET at weekly intervals as soon as the trees bloom. If substantial rainfall occurs, adjust ET by subtracting rainfall (about 70% of total) from accumulated ET. The first irrigation is needed during the week of April 12, as the cumulative ET approaches the allowable depletion of 4.0 inches (10 cm). Schedule the second irrigation for about May 8, when another 4.0 inches are about to be depleted. Apply 5.3 inches (13.5 cm), the gross irrigation requirement, at each irrigation.

cover, you will need to apply additional nitrogen for use by the cover plants.

Other nutrients that may be deficient in almond orchards are zinc, potassium, copper, iron, and boron. Zinc deficiencies are fairly common in sandy soils or young, drip-irrigated orchards and should be treated yearly once a deficiency has been confirmed. Potassium deficiencies are found in some sandy soils of the San Joaquin Valley and in some finer-textured soils in the northern Sacramento Valley. Copper and iron deficiencies are more limited in occurrence. Boron deficiencies occur in a few areas in the northern San Joaquin Valley and Sacramento Valley. Magnesium, manganese, and phosphorus deficiencies are rare in almonds.

The tree's ability to take up nutrients depends on the type of fertilizer, soil moisture, soil type, and temperature. Make sure you apply the appropriate fertilizer at the right time and in the right place. For information on application methods, rates of application, materials, and timing to correct various nutrient deficiencies, see *Almond Orchard Management* and *Almond Production*—both listed under General References.

Testing Nutrient Levels. Sample orchards on a regular basis either yearly or every other year to determine nutrient levels. Also, sample nutrient levels if you suspect a problem is the result of a nutrient deficiency or toxicity.

Some nutrient deficiencies produce characteristic leaf symptoms (see *Environmentally-Caused Diseases*, page 122). Confirm suspected deficiencies by having a professional laboratory analyze leaf samples from trees exhibiting symptoms. A directory of commercial labs equipped for nutrient analysis is listed in the Reference section at the back of this manual.

Table 4. Critical Nutrient Levels[1] in Almond Leaves.

Nutrient	Deficient below	Adequate	Adequate over	Toxic above
nitrogen	2.0%	2.2–2.7%	. . .	. . .
potassium	1.0%	. . .	1.4%	. . .
zinc	. . .	. . .	18ppm	. . .
copper	. . .	. . .	4ppm	. . .
boron	25ppm	30–65ppm	. . .	85ppm
sodium	. . .	. . .	. . .	0.25%
chlorine	. . .	. . .	. . .	0.3%
manganese	. . .	. . .	20ppm	. . .
magnesium	. . .	. . .	0.25%	. . .
calcium	. . .	. . .	2.0%	. . .
phosphorus	. . .	0.1–0.3%	. . .	. . .

1. Determined by leaf analysis performed in July; percent or parts per million (ppm) based on dry weight.

On bearing trees, sample mature leaves on nonfruiting spurs every year in late June or July to detect deficiencies or possible toxicities. On young nonbearing trees, sample basal shoot leaves. Table 4 lists the critical levels of nutrients that are most often deficient or present at toxic levels. See *Soil and Plant Tissue Testing* listed under Horticulture in References for more details on how to test for nutrient levels.

If a leaf analysis reveals toxic levels of elements, such as sodium, boron, or chlorine, have the soil or water analyzed to determine where these toxic levels are coming from.

Ground Covers

Ground covers are frequently maintained in almond orchards, especially those growing in heavy, fine-textured soils. A ground cover often improves water penetration and aeration of the soil, two important factors for a healthy root system. In addition, a well-managed ground cover provides cooling and reduces dust, which often contributes to mite outbreaks. With a perennial cover, the cooling effect of the ground cover can be advantageous in areas with hot summers but a disadvantage in areas with frost hazards.

A ground cover may consist of volunteer annual plants or a planted annual or perennial crop. Low growing winter annuals that dry up in the summer heat and decompose by harvest make good ground cover plants. If the volunteer ground cover is not adequate, however, it may be necessary to plant an annual fall-seeded crop. Winter annuals such as mustard, wild radish and fiddleneck, which bloom at the same time as almonds, are not desirable cover plants because they may be preferred by bees. To avoid competition by flowering weeds, mow the ground cover just before almond bloom.

A ground cover may be important in providing food and shelter for many predators and parasites of insects and mites, but it also may host certain pests such as disease pathogens and nematodes. Research on the interaction between different types of ground cover and beneficial and pest species is currently underway to determine how vegetation management can be used to its full extent in an IPM program.

To maintain a ground cover properly, be sure to mow frequently enough to avoid trash buildup at harvest. Keep plant growth away from the tree trunks, where it provides conditions favorable for crown rot infections. An orchard with a ground cover uses more water and nutrients than a clean-cultivated orchard; the cost of adjusting irrigation and fertilization programs to account for this extra use must be considered. See the weed chapter for details of vegetation management methods.

Harvesting

Harvesttime varies according to district, season, and variety, but it usually begins in August and extends into October. Table 5 lists early, early-mid, mid-, late and very late harvesting categories of almond cultivars. If trees are knocked too early, nuts will not easily fall from the tree and will require hard knocking to remove them. Hard knocking can result in tree injury, which may lead to Ceratocystis canker. If nuts are allowed to hang on the tree too long, they tend to stick and will not knock easily. Also, nuts allowed to remain on the trees may sustain a higher infestation of navel orangeworm, because they are exposed for a longer period to ovipositing moths. Nuts may also be difficult to remove from trees stressed by lack of moisture or excess mite damage, and the nuts that are removed may be difficult to hull.

Table 5. Harvesttimes of Different Almond Cultivars.

Early	Early-mid	Mid	Late	Very late
Jeffries	Harvey	Carmel	Butte	Drake
Nonpareil	Milow	Carrion	LeGrand	Fritz
	Mono	Jordanolo	Livingston	Mission
	Norman	Monarch	Merced	Planada
	Peerless	Ne Plus Ultra	Monterey	
	Price	Vesta	Padre	
	Sauret #1	Yosemite	Ruby	
	Solano		Sauret #2	
	Sonora		Thompson	
	Tokyo			

Inspect your trees to determine when to start harvesting. When 95 to 100% of the almonds at eye level have hulls that are visibly split (Stage 3 or more, see Figure 6, page 12), shake a few trees to determine if nut removal is satisfactory. If too many nuts remain in the tree, try again in a few days. Almonds will still be green at harvest but will dry rapidly on the ground without any loss in nut quality. These nuts are also relatively easy to hull. Nuts that are removed from the tree before splitting, however, will usually dry in such a way that hull removal is not possible.

The orchard floor must be properly prepared before harvest to allow for rapid and efficient nut pickup after the trees are shaken. The ground should be smooth, dry, and free of weeds, ants, trash, and other debris. Orchards that are regularly cultivated should be disced to break up clods and then rolled to make the surface smooth and firm. Some growers apply their last preharvest irrigation after smoothing the soil. Orchards where the soil is not regularly cultivated, but where the weeds are mowed or managed with herbicides, require fewer preharvest operations. Weeds growing between the tree rows should be flail-mowed close to the ground so that they will sufficiently decompose before harvest. With either ground management system, the last preharvest irrigation should be 1 to 4 weeks before harvest, depending on soil texture and depth. To avoid germinating a new crop of weeds, you may add a preemergence herbicide to the water.

Harvest at the earliest possible date to avoid worm damage to nuts and to minimize damage from hull rot and early rains. If danger of a navel orangeworm infestation is severe, it may be cost effective to harvest before maximum nut removal is possible. Nuts left behind can be harvested with other cultivars.

After nuts are knocked onto the ground, the hulls are allowed to dry before the nuts are swept into windrows and picked up. During this period they are susceptible to damage by ants; the longer they remain on the ground, the greater the potential for damage. If an economically damaging number of ants are present in orchards, they should be controlled before harvest (see the insect and mite section for further details).

Unless nuts can be hulled rapidly and delivered to the handler, consider on-farm fumigation immediately after harvest to stop the amount of damage to the crop by navel orangeworm and peach twig borer.

Pruning

Nonbearing trees are pruned in order to shape and train the limbs for structural strength while developing maximum fruiting area and fruitfulness. Bearing trees are pruned to renew the fruit wood for consistently high yields from year to year. In addition, diseased, weakened and broken limbs, interfering branches, and unneeded water sprouts are removed.

Almond trees bear their fruit laterally either on spurs that live about 2 to 5 years or, depending on vigor and cultivar, on long shoots. Prune about one-fifth of the fruiting wood each year to stimulate adequate shoot growth. The actual amount of pruning, however, depends on cultivar, fertilization, tree vigor, and previous crop. When large limbs weaken, die, or break, prune these limbs flush at the base.

Sanitation

Sanitation is an important management tool in almond orchards because of the long life of the trees. Pests such as perennial weeds, nematodes and soil-borne disease pathogens are difficult to control after trees are planted. Prevent introduction or spread of infestations by these and other pests by following the precautions listed below. The most important of all sanitation practices is the removal of mummy nuts from the orchard following harvest.

- Shake mummy nuts from the trees, and remove or destroy trash nuts to deprive the navel orangeworm of overwintering sites.
- Clean equipment used in other areas to avoid introducing pathogens, nematodes, or weeds into the orchard.
- Keep tree crowns free of weeds to reduce crown rot infections and mice infestations.
- Spot treat infestations of perennial weeds as soon as they are discovered in the orchard to prevent them from spreading throughout the orchard.
- Don't irrigate with tailwater from fields infested with nematodes or other soil-borne pathogens, and don't use water that may carry harmful herbicide runoff.
- Destroy stands of weeds along the borders of the orchard to reduce mice infestation. Weeds allowed to mature around an orchard also produce seeds that infest the orchard.
- Remove roosting, nesting, or resting areas, when possible, that might attract large numbers of crop-destroying birds.
- Remove stumps, brush piles, and debris to limit ground squirrel populations and prevent population buildups of wood-boring insect pests.

Biological Control

Potential pest populations in the almond orchard are frequently controlled by other organisms inhabiting the orchard. These biological control agents include predators, parasites, pathogenic microorganisms, and competitors. They may be native or introduced from other areas. The importance of biological control in almonds is best known for mites. Natural enemies are also important in the control of scales and certain caterpillars. However, biological control is not reliable for the two key insect pests, navel orangeworm and peach twig borer.

Biological control may be disrupted by chemical treatments for key pests, resulting in outbreaks of mites or scales. You can reduce such secondary outbreaks by carefully selecting types and rates of insecticides and timing applications so that they are least disruptive to the natural enemies.

Biological control of diseases and weeds is very limited at present. One almond disease, crown gall, can be controlled on young trees at planting time by dipping them in a solution of a nonpathogenic strain of the crown gall organism (Galltrol). The nonpathogenic strain prevents the pathogenic strain from infecting the roots by out-competing it. Biological control of puncturevine by a stem and seed weevil may occur in nontilled orchards. Where tillage occurs, however, control is disrupted. Weevil populations may build up in puncturevine growing in nontilled areas adjacent to the orchard, but not sufficiently to give control in a tilled orchard.

Pesticides

Within an IPM program, pesticides are valuable tools because they can reduce pest populations drastically within a short time. In many cases, they are the only control tools available.

Use of Pesticides. Pesticides, including insecticides, herbicides, fungicides, and rodenticides, belong to a variety of chemical groups with very different properties. Often you can choose among several chemicals and different formulations to control one pest. On the other hand, you may be able to select one chemical to control several pests at the same time or use compatible materials together. The choice will depend on the degree of control needed, the effect on other pests and beneficial species, and on economic and legal restrictions. Use the following guidelines to select and use pesticides:

1. Where possible, choose the least disruptive chemical. Some synthetic organic materials used for control of the navel orangeworm, for example, are more disruptive to natural enemies of mites than others.
2. Treat for insects and mites only when monitoring indicates that economically damaging populations are present or anticipated. Try to apply the pesticide when the target pest is at its most vulnerable state and when natural enemies are least affected. For example, spraying during tree dormancy is very effective for certain insect pests and has little effect on natural enemies.
3. When treating weeds after emergence, be sure they are at the growth stage recommended on the label.
4. Calibrate the spray application equipment properly, use the correct dosages, and, where possible, place the chemical selectively for most efficient control.
5. Consult your farm advisor for the lowest effective dosage and best application method for the pest problem in your orchard. Recommended dosages and frequency of application given on pesticide labels may be higher than necessary for good control when properly timed and applied. Recommended dosages for herbicides usually have to be adjusted for soil type, climatic conditions, and irrigation method used. More information on applying insecticides is given in the insect chapter of this manual.
6. Before choosing a pesticide, check the latest University of California recommendations for almonds. Always read the pesticide label; it outlines safety precautions, legal requirements, and registered uses for each product.

Ground Application. Nearly all ground spraying for insects, mites, and pathogens in commercial almond orchards is done with air carrier (air blast) sprayers (Figure

Figure 16. The air blast sprayer is the most commonly used applicator for ground applications of pesticides in almond orchards.

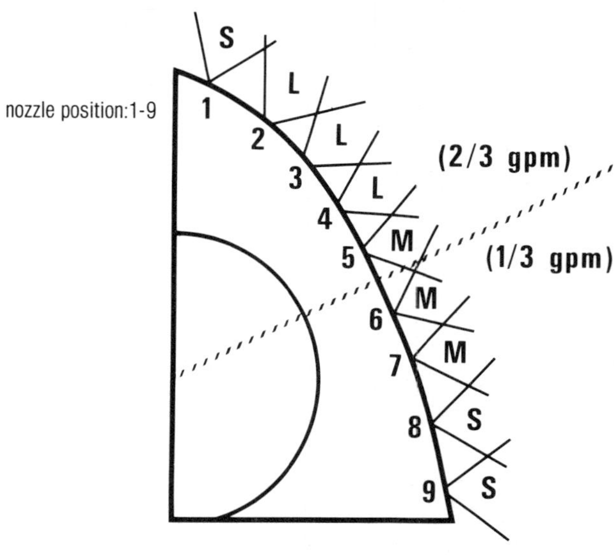

Figure 17. Nozzle arrangement in an air carrier sprayer. Arrange nozzles so that two-thirds of the spray discharge in gallons per minute (gpm) comes out of the top one-half of the sprayer manifold. Relative size of nozzles: L = large, M = medium, S = small.

16). (For information on herbicide application, see the weed section of this manual and *Calibration of Herbicide Sprayers* in References.) In most air carrier sprayers the pressurized, pesticide-water mixture is pumped to a series of nozzles that are directed into a blast of air from a fan that blows the spray into the tree. Air carrier sprayers produce relatively small to large volumes of air (4,000 to over 100,000 cubic feet per minute). Moving at 80 to 180 miles per hour as it passes through the nozzle, the air picks up the pesticide spray and distributes it throughout the tree.

The interior of the tree contains still air. As the sprayer passes each tree, the spray-laden air from the sprayer must mix with the tree air. If the sprayer passes by too fast, or if the tree is too dense and the sprayer output too small, complete mixing with the air in the tree does not take place. The result is poor coverage and control. The ideal air carrier is one that can uniformly and effectively distribute spray materials over all tree surfaces.

Nozzles in an air carrier sprayer are usually mounted in a semicircular manner either in the air stream of the sprayer or adjacent to it. Most machines have one bank (manifold) of nozzles on each side, but two banks are sometimes used. To select and arrange nozzles, use the spray manual or manufacturer's nozzle chart. Nozzles should be arranged so that two-thirds of the spray discharge comes out of the top one-half of the sprayer manifold (Figure 17). The same ratio of air discharge is also necessary. This arrangement prevents overspraying the bottom portion of the tree at the expense of the top portion, where most of the persistent pest problems occur.

When nozzles wear out, they no longer reliably emit the proper droplet size or rate of discharge. Check your nozzles frequently; wettable powders are more abrasive than liquid chemicals and tend to wear out nozzles more rapidly. In dilute spraying, hollow-cone style and the so-called nonplugging type of nozzle are used. These nozzles are made from common steel and brass as well as nylon, stainless steel, and ceramic materials. In conventional concentrate and semi-concentrate spraying, only materials such as stainless steel, ceramic, or tungsten carbide should be used because of the abrasive nozzle wear. Nonconventional, low-volume air shear nozzles, which operate at low pressures, are not subject to serious wear.

Air carrier sprayer tanks (which constitute most of the machine) are usually made of coated or stainless steel, or in some cases chemical-resistant plastics or fiberglass. Stainless steel or fiberglass tanks are preferable. Strainers in the tank and the hydraulic system are necessary to separate foreign materials to prevent clogging. Mechanical tank agitation is used in most sprayers, although hydraulic agitation is used in some makes.

Coverage. Complete coverage of the tree is essential for maximum control with a pesticide. All parts of the tree that may harbor the pest—trunk, limbs, leaves, and fruit—must be reached. Two major causes of poor coverage are improper nozzle selection and arrangement (Figure 17) and incorrect ground speed.

Ground application equipment should not be moved through the orchard faster the 2.5 mph for dilute spraying and 2.0 mph for concentrate spraying. If you go faster, the spray will not adequately cover certain parts of the tree. Speeds even slower than these are recommended if the trees are dense or have a bad scale infestation. To maintain proper ground speed, use a tractor speedometer, or guage your speed by timing your travel over a set distance or by counting the number of tree spaces you pass in a minute (Figure 18).

Gallonage. Pesticides are applied either in concentrate (low-volume or no drip) sprays or in dilute (high-volume or runoff) sprays. Concentrate sprays use smaller quantities of water than dilute sprays and produce greater numbers of droplets that are smaller and more uniform in size than dilute sprays (Figure 19). Generally, concentrate sprays use 25 to 100 gallons per acre and discharge about 2 to 5 gallons of spray per minute from each side. High-volume sprays use 350 to 800 gallons per acre and discharge 12 to 40 gallons per minute from each side of the manifold. The exact amount to use depends on tree density and size, economics, previous experience, and personal choice.

Concentrate sprays are as effective as dilute sprays, but require a greater degree of precision during application. Concentrate spraying has several advantages over dilute spraying, including the use of smaller and possibly less costly equipment that is cheaper to maintain. This method uses less water (70 to 90% less) and therefore requires fewer tank fills during treatment of an orchard. Concentrate spraying can also use 25 to 30% less chemical per acre because the spray does not run off the tree as it does with dilute sprays. Small spray droplets from properly calibrated concentrate applications stick to tree surfaces on contact.

Presently, three types of concentrate sprayers are in use in California tree fruit and nut crops. The most common type is the older sprayers that were basically designed for dilute application, but have been modified for concentrate spraying through changes in pump pressure, spray manifold, and nozzle number, size, and placement. Newer models of these sprayers are easier adapted to concentrate spraying because they have 100 pounds per square inch at the manifold, 100 to 160 miles per hour air velocity, and a second manifold arrangement with smaller hollow cone, disc-type nozzles for concentrate sprays.

The third group is sprayers that were basically designed for concentrate application. Some of these use the same kind of nozzles, pressure, and air velocity systems as do the dilute application models. Others use a low-pressure system to move spray liquid to special

1. Calculate how many feet you travel in one minute.
(This example is for a desired speed of 2 MPH.)

$$\text{Feet/min.} = \frac{\text{mph} \times \text{ft/mile}}{\text{min/hr}} = \frac{2.0 \times 5280}{60} = 176 \text{ Ft.}$$

2. Calculate how many tree spacings you should pass in a minute.
This example uses 22 ft. tree spacings.

$$\frac{176 \text{ Ft.}}{22 \text{ Ft./row}} = 8 \text{ rows}$$

If it takes you more or less than one minute to pass 8 tree spaces, adjust your ground speed.

Figure 18. Worksheet for checking groundspeed.

Figure 19. Diagrammatic drawing shows how coverage depends largely on the droplet size of the spray. The same amount of spray from a concentrate sprayer can cover a much larger area than that from a dilute sprayer. For each 300-micron droplet formed by a dilute sprayer, a concentrate sprayer can discharge 216 (6 x 6 x 6) droplets of 50 microns. (As illustrated, each droplet is surrounded by a 100-micron zone, in which each spray droplet is effective.)

a. **List the known criteria of your operation:**

1. row spacing (trunk to trunk, ft.) 22
2. gallons of spray per acre (gpa) 100
3. desired ground speed (mph) 2.0

b. **To find the gallons per minute (gpm) for the nozzle manifold (one side), use the following equation:**

$$\text{gpm} = \frac{\text{gpa} \times \text{mph} \times \text{row spacing}}{1000}$$

$$\text{gpm} = \frac{100 \times 2.0 \times 22}{1000} = 4.4$$

4.4 gallons per minute need to be discharged from one side (8.8 gpm from both sides) to deliver 100 gpa at 2.0 mph.

Two-thirds (2.9 gpm) should be discharged from the top ½ of the manifold.

One-third (1.5 gpm) should be discharged from the bottom ½ of the manifold.

1) The equation is derived from:

$$\text{gpm} = \frac{(\text{gpa})}{2 \text{ sides}} \div \text{min./A.}$$

$$= \frac{\text{gpa}}{2} \div \frac{\text{sq. ft./A.}}{\text{ft./min} \times \text{row spacing}}$$

$$= \frac{\text{gpa}}{2} \times \frac{\text{ft./min.} \times \text{row spacing}}{\text{sq. ft./A.}}$$

$$= \frac{\text{gpa}}{2} \times \frac{\text{mph} \times \text{ft./mi.}}{60 \text{ min.}} \times \frac{\text{row spacing}}{\text{sq. ft./A}}$$

$$= \frac{\text{gpa}}{2} \times \frac{\text{mph} \times 5280}{60} \times \frac{\text{row spacing}}{43560}$$

$$= \frac{\text{gpa} \times \text{mph} \times 88 \times \text{row spacing}}{2 \times 43.56 \times 1000}$$

$$= \frac{\text{gpa} \times \text{mph} \times \text{row spacing}}{1000}$$

Figure 20. Worksheet for calibrating air carrier sprayers.

nozzles where the air-discharge shears spray into small droplets. Regardless of nozzle system used, a relatively narrow range of small droplet sizes is preferable.

Use the formulas in Figure 20 to calculate the number of gallons per minute needed from each side of the nozzle manifold for either a dilute or concentrate spray.

Aerial Application. Aircraft applications in orchards are usually not as satisfactory as ground applications are for insect and mite control after trees are in full foliage. Air applications do not cover the lower area of the tree or the undersides of the leaves as well as properly applied ground sprays. Air applications, however, are useful for dormant and delayed-dormant sprays and for fungicide treatments in early spring when the ground is so wet that spray equipment would cause serious wheel ruts and soil compaction. Aircraft applications may also be needed at times to cover large areas in a very short time such as in the case of navel orangeworm sprays at hull split. For best coverage, apply aerial sprays by helicopter at not less than 25 gallons per acre and at 30 miles per hour. Because the first almonds to split are in the upper, outer portions of the tree canopy, a properly applied aerial application can provide effective control against the navel orangeworm at this time. If you are spraying for mites at the same time, however, aerial sprays may not give adequate mite control, particularly if mite predators are not present in adequate numbers.

Problems Associated with Pesticides. Attributes that make pesticides valuable control tools may in some circumstances bring problems that directly influence the success of an IPM program. These problems include development of pest resurgence, secondary pest outbreaks, and resistance. Pesticide applications may also affect neighboring crops, wildlife, and human health.

Pest Resurgence. Pest resurgence refers to the rapid rebuilding of a pest population that was recently controlled by a pesticide application. Such resurgence is especially a problem in pest populations that have short reproductive cycles, such as mites. Faced with resurging populations, you may have to spray repeatedly, increasing your production costs and chances for creating secondary pest outbreaks.

Secondary Pest Outbreak. Secondary pest outbreak refers to the buildup of insects and mites that were previously controlled by natural enemies. In almonds, treating for navel orangeworm with certain materials can cause outbreaks of mites by killing mite predators but not the pest mites. To reduce chances for pest resurgence and secondary pest outbreak, choose selective materials when possible, and time applications to be most destructive to pest species but least destructive to beneficial species.

Resistance. Development of resistance to pesticides can occur in pest populations, particularly in those that reproduce rapidly, such as pathogens and mites. Although less common, resistance has also been reported in some rodent and weed species. To avoid resistance, it is desirable to alternate materials that have a different mode of action when these are available. To slow the development of resistance, use pesticides only when needed and apply the materials selectively.

Hazards to Bees. Many pesticides used in almonds are toxic or otherwise harmful to bees. Bees must be protected from pesticides during the time that they are in the orchard. If you are applying a delayed dormant spray, be sure to observe the minimum interval between application of the insecticide and placement of bees in the orchard. Do not use any pesticides, even those considered non-toxic to adult bees, during bloom unless absolutely necessary. Bees foraging in almond flowers treated with certain pesticides, including some fungicides, can return to the colony carrying enough chemical to damage the colony. Chemicals may not affect adult bees, but larvae exposed to them may die or develop into malformed adults.

Take the following precautions when spraying orchards in the spring: notify the beekeeper if you must spray, and apply sprays in early morning or late evening when bees are not active; don't spray over hives or allow pesticides to get into hive entrances; and avoid drift of dormant sprays onto broadleaved winter annuals if bee hives are within 2 miles of the orchard. Because bees often forage 2 to 3 miles from their hives, a community effort is important to protect bees from pesticide damage. For details of toxicity categories of pesticides and spray precautions, see *Reducing Pesticide Hazards to Honey Bees with Integrated Management Strategies* listed under Pesticides in References. Consult your agricultural commissioner for legal restrictions.

Hazards to People and Wildlife. Many of the pesticides used in almonds are potentially hazardous to people and wildlife. Applicators are most heavily exposed to pesticides, and they may be at risk of acute poisoning or chronic ailments if adequate safety procedures are not followed. Minimize health problems by following label directions, wearing appropriate protective clothing, observing state and local regulations, including reentry intervals, and confirming the availability of emergency health care. To reduce drift, check the wind speed and direction and the calibration of the equipment. Avoid drift into ponds, rivers and other water supplies.

Vertebrates

Figure 21. Ground squirrels can damage branches by gnawing on the bark and occasionally girdling a limb.

Almond orchards provide food and shelter for a variety of vertebrate pests. The most damaging of these pests is the ground squirrel, which causes losses by feeding on nuts and, to a lesser degree, leaves and bark. Ground squirrel feeding may girdle a young tree, or ground squirrels may cause problems by gnawing on drip lines or by digging burrows in the orchards. Burrows, in turn, may expose roots, disrupt irrigation, and interfere with mechanical harvesting. Crows, scrub jays, and magpies can also cause substantial losses by consuming or carrying off large numbers of nuts. Pocket gophers present problems in the orchard by girdling young trees and feeding on roots of older trees. In almond orchards located adjacent to wooded areas, eastern fox and eastern gray tree squirrels may take large amounts of nuts. Deer, meadow mice, rabbits, and coyotes may occasionally damage trees or irrigation lines in some growing districts.

Location of an almond orchard and the surrounding habitat determine to a large extent which vertebrate species may be involved in orchard or crop damage. For example, small orchards in the foothills or adjacent to rangeland, wooded areas, or stream banks are likely to experience ground squirrel problems. Large, extensive plantings in valley locations, especially flood-irrigated orchards where the surrounding land is mostly cultivated, may have fewer ground squirrels. The potential for vertebrate damage can also be influenced, to a lesser extent, by management practices in the orchard. Flood irrigation sometimes discourages infestation by meadow mice, pocket gophers, and ground squirrels. Tillage or total reliance on herbicides may also reduce some vertebrate problems by eliminating weed growth that might shelter certain vertebrate pests, such as meadow mice.

Predators, diseases, and available food sources all influence a vertebrate population. Predators, such as foxes, badgers, snakes, hawks, and owls feed on many vertebrate species, but they cannot be relied on to prevent vertebrates from causing economic damage to crops. Such predators should be encouraged, however, where they do not pose a hazard to farm animals.

A few preventive measures can make the orchard or the surrounding areas less favorable to pest invasion and survival of some vertebrate pest species. For example, removing debris, brush or prunings in or around the orchard will deprive ground squirrels of favored cover. Total weed control or mowing may reduce meadow mice infestations in the orchard. Or, clearing weeds around young tree trunks may discourage girdling by meadow mice. If the preventive approach proves unsatisfactory or unfeasible, other management methods, such as traps, toxic baits or fumigants, may have to be used.

Table 6. Legal Classification, Classifying Agency, and Legal Control Restrictions of Vertebrate Pests.

Species	Legal status and restrictions
Ground squirrels	**Nongame mammals: California Fish and Game Code.** Nongame mammals found injuring growing crops or other property may be taken at any time or in any manner by owner or tenant of premises.
Crows and magpies	**Migratory nongame birds: Code of Federal Regulations.** Federal permit not required to control crows and magpies when found depredating or about to depredate a crop.
Scrub jays	**Migratory nongame birds: Code of Federal Regulations.** May be controlled only under a depredation permit from the U.S. Fish and Wildlife Service.
Tree squirrels	**Game mammals: California Fish and Game Code.** Permit from the regional office or local game warden is required for control of eastern gray and western gray squirrels. Permit is not required for eastern fox squirrel if it is depredating a crop. Tree squirrels may be shot or trapped only; poisoning of tree squirrels is illegal.
Deer	**Game mammals: California Fish and Game Code.** When deer are found to be damaging crops, a permit must be obtained from local game wardens to control deer by shooting. Growers must adhere to a variety of permit regulations, including field dressing and delivery of animal carcasses as designated by the game warden. Deer may also be hunted during the regular deer season following all appropriate regulations. All other methods of destroying deer are illegal.
Black-tailed jackrabbits	**Game mammals: California Fish and Game Code.** Hunting season is all year with no limit. When injuring crops or other property, jackrabbits may be taken at any time or in any manner by owner or tenant of premises.
Brush and cottontail rabbits	**Game mammals: California Fish and Game Code.** Observe seasons and bag limits for sport hunting. Brush and cottontail rabbits may be killed by shooting or trapping at any time if injuring a crop. Poisoning brush and cottontail rabbits is illegal.
Coyote	**Predators: California Fish and Game Code.** Coyotes can be taken when found injuring property. Contact county agricultural commissioner for the availability of federal, state, and county control programs in your area.

State and federal regulations concerning control methods differ according to species (Table 6). Control materials that may be used for a particular species may vary from county to county. Consult your county agricultural commissioner for information on how pests can best be controlled in your area.

Because each vertebrate pest must be managed differently, it is important to correctly identify the species that is causing the damage. In the orchard and in adjacent areas, watch for the animals or indications of their presence, such as burrows, soil mounds, feces, nut shells, gnawing marks on the bark, or damage to irrigation lines.

Timing of a control action is often critical. For instance, the optimum strategy for controlling ground squirrels is to fumigate after they have emerged from hibernation in late winter but before the young are active aboveground in late spring. Fumigation is best at this time because the soil is moist and retains the toxic gas well. Cereal baits generally are not well accepted and do not provide good control in spring because squirrels are feeding mainly on fresh vegetation and are not attracted to seeds.

Because vertebrate pest problems are complex, no guidelines are available that specify what infestation levels and combination of environmental conditions justify control actions. Use the biological information presented in this chapter, the history of vertebrate pests in your orchard, and your knowledge about the cost effectiveness of available control methods to determine your course of action.

When controlling a particular vertebrate pest, treat all surrounding infested areas including roadside ditches, adjacent fields, or wooded areas. Also, try to coordinate your vertebrate control program with neighboring growers to increase effectiveness of the program.

After you have achieved control, establish a routine monitoring program to detect reinvasions and to make future control decisions easier and more efficient. For more information on vertebrate control problems, see the section on Vertebrates in References.

Ground Squirrels
Spermophilus beecheyi

Ground squirrels are the most damaging vertebrate pests in almonds. They are especially destructive in almond orchards located adjacent to rangelands. In addition to being near other favorable squirrel habitat, these orchards are frequently sprinkler or drip irrigated. Flood-irrigated orchards tend to have fewer squirrel problems because flooding discourages squirrels from digging burrows directly in the orchard, although they may simply locate their burrows on higher ground at the orchard border.

Ground squirrels may feed on blossoms and buds, but the greatest loss is to developing and mature nuts. Ground squirrels can climb the trees and strip branches of large numbers of nuts. They also occasionally gnaw on the bark and can girdle the trunk or scaffolds (Figure 21). They can also damage plastic irrigation lines. Their burrows may disrupt irrigation and can cause erosion; burrows can also interfere with harvest operations. In some areas, ground squirrels may present a potential health problem because they can carry human pathogens, such as the one that causes bubonic plague.

Several subspecies of the California ground squirrel are found in the almond-growing areas of the state, but in any one locality only one subspecies is present. Adult squirrels weigh 1 to 1.5 pounds (0.5 to 0.7 kg). They are about the same size as tree squirrels, but their tails are not as bushy and their fur is mottled with flecks or a dark patch on the back (Figure 22). Ground squirrels often bask in warmth of the morning or late afternoon sun. The are social animals, living in colonies that sometimes get quite large.

When ground squirrels themselves are not seen, their burrows (Figure 23) are a good indication of their presence. Each burrow system can have several openings with scattered soil in front. In contrast, burrows of pocket gophers have plugged entrances. Individual ground squirrel burrows may be 5 to 30 feet (1.5 to 9 m) long, 2.5 to 4 feet (75 cm to 1.2 m) below the surface, and about 4 to 6 inches (10 to 15 cm) wide. Burrows provide a place to retreat, sleep, hibernate, rear the young, and store food.

Ground squirrels often dig their burrows along ditches, fence rows, and on uncultivated land (Figure 24). Where orchards are not flood irrigated, squirrels also move into the orchard and dig their burrows beneath the trees.

Figure 25 shows the major activity periods and food sources of ground squirrels. They emerge from hibernation during late January in the southern part of the Cen-

Figure 22. The California ground squirrel has mottled brown to gray fur with flecks or a dark patch on its back.

Figure 23. Ground squirrels may locate their burrows directly in the orchard. Note that the squirrels have been gnawing on the bark of the tree trunk.

tral Valley and up to two months later in the northern part. Females have one litter of six to eight young in the spring. About 6 weeks after birth, the young emerge to feed aboveground. Ground squirrels feed on green vegetation in the spring and early summer. As vegetation dries up in the early summer, they switch to seeds, fruits, and nuts.

Adults often go into a temporary state of inactivity (aestivation) for part of the hot summer and into hibernation in the winter. Some young may not aestivate or hibernate during the first year. Because of these periods of inactivity, squirrels seem to be more numerous in the spring, early summer, and fall. Ground squirrels can live 5 years or more.

Management Guidelines

The need to take action against the ground squirrel largely depends on local conditions. If your orchard is surrounded by infested, uncultivated lands, reinvasion will be a perennial problem. In cultivated areas where only a limited amount of highly desirable habitat exists, removing stumps, brush piles, and debris can be helpful in limiting populations or at least in allowing you to detect them more readily. Try to prevent the initial establishment of ground squirrels in your orchard. Once ground squirrels have become established, the orchard is likely to be reinvaded, because old burrows are attractive to future populations. Filling in the tops of these burrows will not reduce the rate of reinvasion.

To monitor ground squirrels, check the perimeter of the orchard about once a month for animals or their burrows. Monitor during mid-morning or late afternoon when squirrels are most active. Record the approximate number of squirrels and the location and number of burrows. You can expect a great increase in ground squirrel activities when the adults emerge from hibernation and again when the young first become active.

If your monitoring records indicate that the squirrel population is living or feeding in the orchard, consider control methods such as traps, fumigants, and toxic bait. Choose the method carefully according to the time of

Figure 24. Ground squirrel burrows may be located around the orchard's edges such as along roadsides.

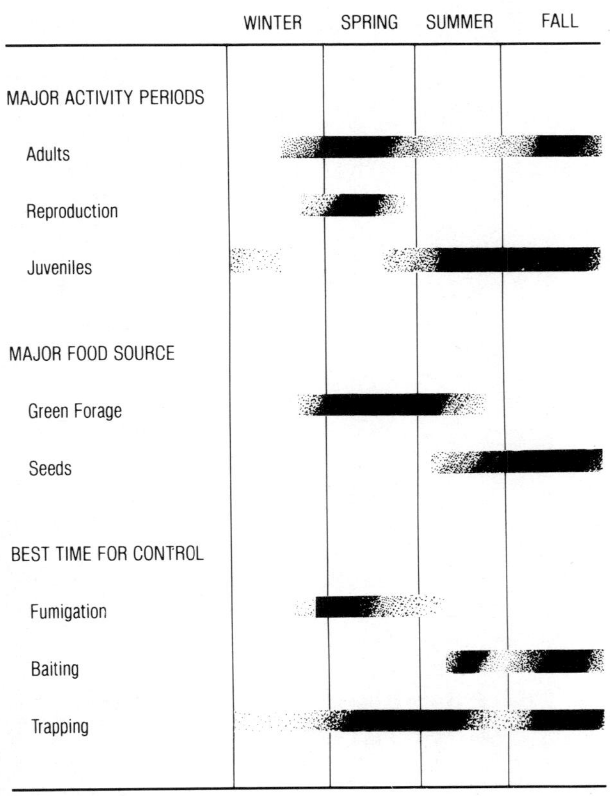

Figure 25. To choose the best time and type of control, consider activity periods and food sources of the ground squirrel. Specific times may vary slightly from region to region, depending on climate.

year (Figure 25), area, and number of squirrels as well as the cost, labor, and potential hazard of the treatment.

You can control ground squirrels most effectively in late winter or early spring after they have emerged from hibernation but before they have reproduced. In timing your control actions, keep in mind that males appear first; all animals have emerged from hibernation when the sex ratio is 1:1. At this time you may either use a fumigant or traps that do not require baiting. Toxic bait provides good control in the early summer when squirrels collect and eat seeds, grains, and nuts. However, as the almonds mature, bait acceptance decreases, and controlling ground squirrels with toxic baits is difficult in late summer and fall. If you are in doubt about correct timing for bait acceptance, consult your agricultural commissioner or farm advisor. Controlling ground squirrels during the hottest part of the summer is not very practical because many of them are aestivating in burrow systems, and young squirrels that remain active at this time of year often prefer maturing nuts to other food sources.

Consider filling in ground squirrel burrow openings following squirrel control. This does not prevent reinvasion by the squirrels, but it is an important step in preparing the orchard floor for harvest and will also help you to detect reinvasion more readily.

Fumigants. You can fumigate ground squirrels whenever soil moisture is adequate, but if squirrels are hibernating, they will not be killed. Fumigants work best in the early spring when moist soil helps retain a high toxic gas level in the burrows and the squirrels have emerged from hibernation. In addition to adequate soil moisture, it is also important that all the active burrows in and around the orchard are treated. When you are treating, double check the orchard and surrounding area to be sure you have not missed any burrows. Retreat opened burrows 3 to 4 days after the initial treatment.

Several types of restricted and nonrestricted fumigants are available. Restricted fumigants include aluminum phosphide, carbon bisulfide, and methyl bromide. These fumigants require a permit from the agricultural commissioner for purchase and must be used by or under the direct supervision of a certified applicator. Aluminum phosphide is manufactured in tablet form. Tablets are placed in the burrow opening, which is then plugged with a crumpled newspaper and sealed with soil. Methyl bromide comes in a pressurized can and is injected as a gas into sealed burrows. Carbon bisulfide is applied as a liquid that rapidly vaporizes in the sealed burrows. Because of its volatile properties, carbon bisulfide must be handled with care. Read the label on all fumigants for application instructions. Some fumigants, such as methyl bromide and carbon bisulfide, are phytotoxic and should not be used in burrow systems that are within the root zones of the trees. Aluminum phosphide is not toxic to almond trees and may be used in burrows that are within the root zones.

Nonrestricted fumigants, such as gas cartridges (woodchuck or smoke bombs) are available from some agricultural commissioners and from several commercial companies. Do not use cartridges where fire may be a hazard, such as near dry grass or beneath a building. When ignited, these cartridges release smoke and toxic gases, in particular, carbon monoxide. Following label instructions, place one or more cartridges in each active burrow opening. Insert and ignite the fuse, push the cartridge deep into the burrow, and quickly seal the entrance. Watch the area for escaping smoke and treat all other burrow openings. Double check the orchard when you are treating to detect any burrow opening you may have missed, and treat these openings as well. Check the burrows after about 3 days. Treat again where opened burrows indicate that squirrels have dug out. Gas cartridges are not toxic to almond trees and can be used to treat burrows beneath the trees.

Traps. Trapping ground squirrels works well in small areas or where there are few squirrels. Live-catch traps are effective but present the problem of disposal of live squirrels.

There are several types of traps that kill ground squirrels. Most types should be placed on the ground near squirrel burrows or runways. Walnuts, almonds, oats, barley, and melon rinds are attractive trap baits in summer and fall.

The Conibear trap is an effective ground squirrel kill trap (Figure 26). The wire trigger permits the trap to be used either baited or, more commonly, without bait. The trap is placed so the squirrel will pass through it, tripping the trigger. It is best to place the Conibear trap directly across the burrow opening. Placing the Conibear trap in a covered box will reduce hazard to children, pets, and poultry. Place bait in the box, but leave the Conibear trap unset for a few days until squirrels enter the box freely.

A box trap for ground squirrels can also be effective (Figure 27). Place the bait well behind or tied to the trigger. Bait the traps but don't set them for several days until squirrels become accustomed to them. After the squirrels are used to taking the bait, rebait and set the traps. Be sure to place traps in such a way that pets or other nontarget animals are not likely to be caught. This trap can be used singly or in groups, enclosed in a larger box. See References for trap suppliers.

Toxic Bait. Poison baits are the control method most commonly used against ground squirrels. Success of a toxic bait program depends on the squirrel's acceptance of the bait. The period of time that ground squirrels will accept bait is sometimes short, especially in the San Joaquin Valley. Therefore, toxic baits may not be the most effective way to control ground squirrels in an almond orchard. Before you put out bait, do some bait-acceptance tests to determine if the squirrels will readily take the bait. Put out a small amount of untreated bait in a few different locations and check several days later to determine if squirrels are feeding on the bait. Willingness of squirrels to feed on offered bait depends to a large degree on time of year and what other food sources are available. Controlling ground squirrels with toxic baits is generally successful as soon as bait acceptance occurs. If your bait tests indicate squirrels are not feeding on the bait, wait and try the tests again later in the season. As nuts reach maturity, however, the chance of bait acceptance decreases because squirrels prefer nuts. You may need to wait until after harvest.

Either single dose or multiple dose baits can be effective, although each is used differently. Multiple dose anticoagulant baits are often used because they are effective and relatively safe for humans and pets. In general, ground squirrels do not become bait shy with anticoagulant baits. Anticoagulants interfere with an animal's bloodclotting mechanism, eventually leading to death. Anticoagulant baits are effective only when consumed in several feedings over a period of 5 or more days. Effectiveness is greatly reduced if 48 hours or more elapse between feedings. Those characteristics, as well as an effective antidote (vitamin K_1), make anticoagulant baits relatively safe. However, care must still be taken to keep bait away from children and pets.

Anticoagulant baits can be used in two ways: in bait stations or by repeated spot (broadcast) baiting. Bait stations (Figure 28) are small structures which the squirrel must enter to eat the bait. Stations contain sufficient bait for repeated feedings and are the preferred baiting method around homes and other areas where children, pets, and poultry are present. Some labels specify the type of bait station that should be used. Entrance holes should be about 4 inches (10 cm) in diameter, large enough to admit squirrels, but not larger animals.

Place bait stations containing 1 to 5 pounds of bait near runways or burrows, and secure them so that they cannot easily be tipped over. If more than one station is needed, space them at intervals of 50 to 200 feet (15 to 60 m). Initially, inspect bait stations daily and add bait as needed. Increase the amount of bait if all is eaten during the day. Fresh bait is important; replace moldy or old bait. It may take a number of days before squirrels become accustomed to the bait station and enter it. Anticoagulant baiting in bait stations generally requires 2 to 4 weeks or more to be effective. It does not immediately affect feeding habits of squirrels. Continue baiting until all feeding ceases and no squirrels are observed. Pick up and dispose of unused bait upon completion of the control program.

If a bait station is not used, repeated spot or broadcast baiting with anticoagulant bait can be effective in controlling ground squirrels. If spot or broadcast baiting is not specified on the product label, however, these baiting methods cannot be used.

Scatter a handful of bait (about 10 placements per pound) evenly over 40 to 50 square feet near active burrows or runways. Re-treat the same location every other

Figure 26. Conibear trap set in burrow entrance. Secure traps with a stake.

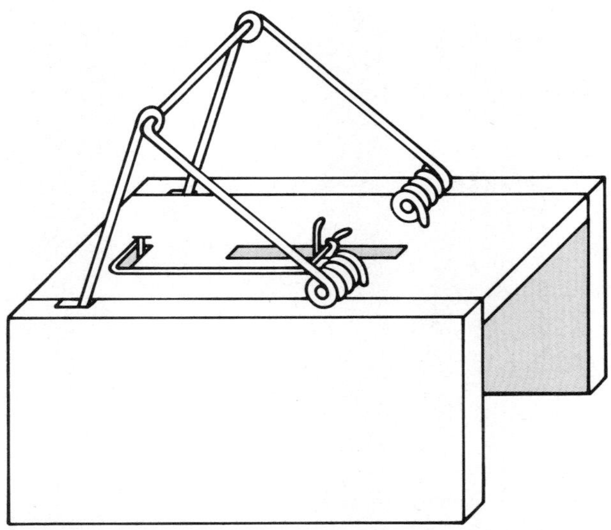

Figure 27. Ground squirrels may be trapped using a box trap.

Figure 28. Unless the bait label specifies otherwise, bait stations can be constructed from durable materials such as metal, wood— or plastic pipe, shown here.

day about three or four times to provide an uninterrupted supply of bait. Scattering bait takes advantage of the squirrel's natural foraging habits. Never pile bait on the ground; it increases the hazard to livestock and other animals which can easily pick up the concentrated bait.

Single dose poisons kill the animals after just one feeding. Most single dose baits, depending on label instructions, can be scattered by hand over 2 to 3 square feet near burrow openings where the vegetation is most sparse. As with anticoagulants, it is important to check bait acceptance with untreated bait before putting out treated bait. Check with your county agricultural commissioner for additional information about individual county programs.

Some anticoagulants and acute toxicants are also formulated to be broadcast by vehicle-mounted seeders or by aircraft. Be sure to read the label to determine if a bait may be applied in this manner. After treating an area, remove and dispose of carcasses if possible to prevent poisoning of dogs or other scavengers that may feed on them.

Figure 29. A single scrub jay can account for considerable crop loss.

Figure 30. Crows damage almonds by feeding on nuts from hull split to harvest, but during the winter they may remove large numbers of mummy nuts that remain on the tree after harvest.

Birds

Birds that can cause significant damage to an almond crop are crows, *Corvus brachyrhynchos*; scrub jays, *Aphelocoma coeralescens*; and in some areas, the yellow-billed magpie, *Pica nuttalli*. Occasionally in the spring, house finches, *Carpodacus mexicanus*, may cause some disbudding in the trees, but this damage generally occurs in isolated areas and is a minor problem statewide.

Crows, scrub jays, and magpies can subsist on many kinds of food, including carrion, but their natural foods are insects, acorns, wild berries, and fruits. Cultivated fruits and nuts are often easy to obtain, and the birds take full advantage of these food sources. Scrub jays (Figure 29) instinctively store food in the fall and carry away more fruit and nuts than they can eat. Thus, a single bird can account for considerable crop loss. Jays are usually solitary birds and occur singly or in pairs; rarely do they group together in flocks. In some orchard situations, with the ready availability of food, however, they have been known to occur in groups. Crows (Figure 30), on the other hand, join into flocks for most of the year and establish a central roost, which can consist of thousands of birds. Each day crows fly from the roost to feeding areas. Magpies generally occur in smaller groups.

Crows, scrub jays, and magpies cause the greatest amount of damage during the period from hull split to harvest. After harvest, these birds may be considered beneficial because they can remove large numbers of mummy nuts left in the orchard.

Management Guidelines

If an orchard is planted in an area that is near attractive resting, hiding, or nesting areas, the crop may be damaged by birds. Features that are particularly attractive to some pest birds include groves of oak or eucalyptus trees, areas along rivers and streams, and orchards planted near powerlines.

When they are depredating or about to depredate a crop, crows can only be controlled by shooting, bow and arrow, falconry, and frightening; however bow and arrow and falconry are not practical control methods in commercial orchards. Magpies may be controlled by shooting or frightening. Shooting may cause the birds to leave an area, but it is time consuming and impractical in large orchards.

The most effective way to reduce scrub jay populations is by shooting; this requires a depredation permit from the U.S. Fish and Wildlife Service. Frightening devices may at times provide some control against scrub jays.

A variety of frightening devices, such as automatic gun exploders, shell crackers, or distress and alarm calls are commercially available for scaring birds. Although bird damage can sometimes be reduced by using only one type of device, frightening devices are most effective when used in various combinations, particularly for crows. These are intelligent birds that are difficult to keep away from orchards. You may get the best control by igniting various automatic frightening devices, while a person moves through the orchard exploding shell crackers or firing a shot gun. Scaring equipment is also more effective when the devices are moved often to prevent birds from becoming accustomed to a single noise pattern. Be sure to have an adequate number of frightening devices; elevate them above the tree canopy so that the sound will not be muffled. Start the frightening regime before the birds establish regular feeding patterns in the orchard. Once feeding habits are established, they are difficult to break using frightening techniques.

Each orchard represents a unique environment and must be evaluated individually to devise an appropriate bird control program. Contact your county agricultural commissioner for assistance and additional information on pest control methods.

Pocket Gophers
Thomomys spp.

Pocket gophers frequently invade almond orchards, especially those with ground covers or sprinkler or drip irrigation; flood irrigation tends to discourage the buildup of a gopher population but does not eliminate them. In an almond orchard, gophers feed mostly on the roots of herbaceous plants and weeds, but they may also feed on the roots of almond trees (Figure 31) or girdle the trunks of young trees just below the soil level. Pocket gopher mounds interfere with mowing or harvesting operations, and the tunnel systems sometimes divert irrigation water.

Pocket gophers are rarely seen, but characteristic plugged mounds of soil indicate their presence (Figure 32). The gophers dig tunnels 8 to 12 inches (20 to 30 cm) below the surface, push the soil out lateral exits and plug these openings (Figure 33). On irrigated land, gophers breed almost throughout the year with a peak in late winter, and the female may bear three litters averaging five to six young.

Management Guidelines

Monitor gophers in late fall, winter, or spring. Gophers increase their digging activity when soil is moistened by rain or irrigation, and you can easily detect fresh mounds. Detection is also easier after mowing when the ground cover is low.

Figure 31. Pocket gophers can damage almond trees by feeding on roots. Some of the roots of the tree in this photo have been eaten by gophers (note the underground tunnel next to the trunk), greatly reducing the upper portion of the root system.

In the spring, you may find new gopher mounds in various parts of the orchard as the young establish burrow systems of their own. Check the orchard about once a month, especially the perimeter where gophers may move in from adjacent fields or orchards.

If your monitoring records indicate that a gopher population is already in the orchard or is expanding, you have two control options: traps or toxic bait. Trapping and baiting can be done anytime of the year, but both methods are easier when the soil is not too dry or hard. Fumigants do not work well because the burrow system is extensive, and gophers can quickly seal their tunnel when they detect the poisonous gas.

Traps are useful especially when the area of infestation is small. Locate the main tunnel, which usually extends from the lower side of the mound, with a 1/4-inch rod or a specially designed metal probe (Figure 34a). Place traps into the main tunnel as shown in Figure 34b. For the most effective control, inspect traps twice daily.

Large populations of pocket gophers are controlled most economically with poison baits. Baits can either be dropped by hand into underground runways or distributed by tractor-drawn mechanical bait applicators. Mechanical bait applicators have greatly improved pocket gopher control by providing a rapid and inexpensive means of eliminating much of the gopher population in infested orchards with a once-over operation.

The mechanical bait applicator constructs an artificial burrow beneath the soil and deposits poisonous grain at fixed intervals. The artificial burrow intercepts most of the gopher burrow systems one or more times, and pocket gophers readily explore the artificial tunnels and eat the bait. Limit the use of mechanical bait applicators to areas of the orchard where gophers are present. If you follow up with hand probing, begin about 2

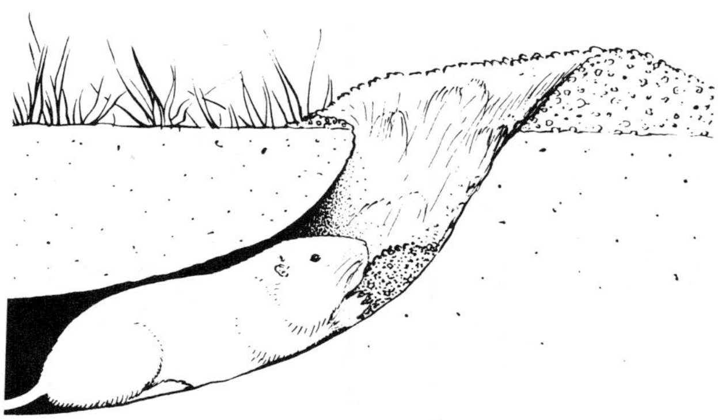

Figure 32. Pocket gophers dig their tunnels by pushing soil out through lateral tunnels. They plug openings with soil after completing the tunnel.

Figure 33. Mounds of loose soil mark the entrance to the plugged tunnel of a pocket gopher.

weeks after mechanical baiting. For more information on mechanical baiting, see *Pocket Gopher Control with Mechanical Bait Applicator* listed under Vertebrates in References.

Hand baiting is generally used on small acreages or where there are only a few gophers. Commercial grain baits are dropped into tunneled runways through a small opening made with a pointed probe (Figure 34a). The probe is used to locate main runways so bait can be placed where gophers will find it. A runway usually runs in a straight line between two mounds at a depth of 8 to 12 inches (20 to 30 cm). Probing should be done around fresh mounds, or between two fresh mounds, as these indicate the most recent presence of gophers. When the runway is located, the probe will drop about 2 inches (5 cm) as it enters the runway. The opening to the runway then should be enlarged by rotating the probe or by using the large end of the probe. Drop bait in the burrow at two or more places in each runway. Cover the probe hole with a clod or rock to keep out light and to prevent dirt from falling on the bait.

Regulations concerning rodent baits may vary from county to county. Check with your county agricultural commissioner about materials available for use in your county. Follow label directions for application.

a

b

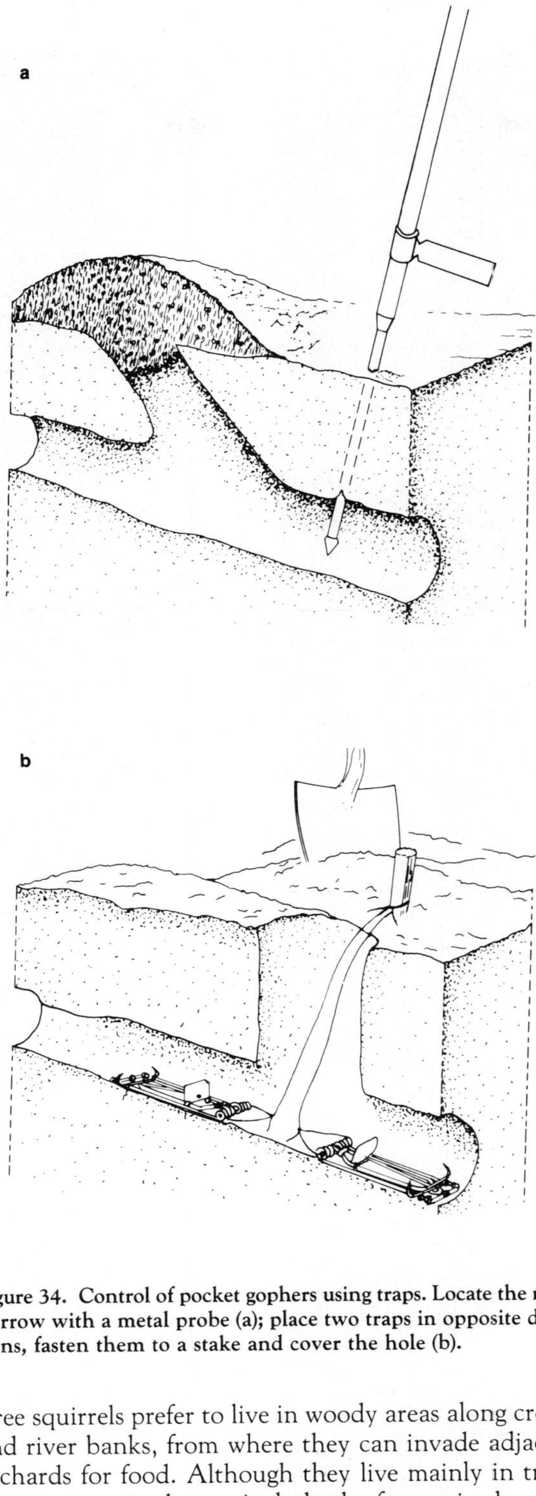

Figure 34. Control of pocket gophers using traps. Locate the main burrow with a metal probe (a); place two traps in opposite directions, fasten them to a stake and cover the hole (b).

Eastern Fox Squirrel
Sciurus niger
Eastern Gray Squirrel
Sciurus carolinensis

In some almond-growing areas, two introduced species of tree squirrels, the eastern fox squirrel (Figure 35), and the eastern gray squirrel, occasionally cause economic damage by feeding on nuts and collecting large quantities for storage. The eastern fox squirrel is much more common than the eastern gray squirrel and is well established in city parks, residential areas, and adjacent agricultural lands, especially in Fresno, Sacramento and coastal areas south of San Francisco. The eastern gray squirrel presently occurs only in a few locations in the San Francisco Bay Area, the San Joaquin Valley, and a small portion of Calaveras County.

Two tree squirrels are native to California: the western gray squirrel, *Sciurus griseus*, which seems to have been largely displaced by the eastern fox squirrel and is rarely a problem to almond growers, and the Douglas tree squirrel, *Tamiasciurus douglasii*, which lives in forests and is not an agricultural pest.

In contrast to the ground squirrel, the tree squirrel has a bushier tail, and its fur is not flecked or mottled. Tree squirrels prefer to live in woody areas along creeks and river banks, from where they can invade adjacent orchards for food. Although they live mainly in trees, some tree squirrels, particularly the fox squirrel, spend considerable time foraging on the ground. The eastern gray squirrel feeds on a variety of green and ripe nuts, seeds, mushrooms, and fruits. The eastern fox squirrel has a similar diet but also feeds on bird eggs and insects. Both

Figure 35. The eastern fox squirrel, more common than the eastern gray, has a bushy tail and reddish brown fur.

Figure 36. To control eastern fox and eastern gray tree squirrels, fasten ground squirrel traps to boards and place them on limbs where damage occurs.

species usually feed only in localized areas of orchards adjacent to wooded areas. The eastern fox squirrel can be distinguished from the eastern gray squirrel by its reddish brown color; the eastern gray looks much like the native western gray squirrel.

Tree squirrels nest in tree holes or build nests of sticks and leaves in the branches of trees. Females of the eastern gray squirrel bear two litters per year of three to five young, one in late winter and one in late summer. Yearling females of the eastern fox squirrel bear one litter; older females have two litters per year between January and April and between July and September.

Trapping is often sufficient to control eastern fox and eastern gray tree squirrels. A ground squirrel box trap can be used for tree squirrels. Fasten one or two box traps on a base board, and nail it on top of a horizontal limb in a tree where you detect feeding damage (Figure 36). Place a handful of nutmeats well behind the trigger mechanism to attract the squirrels. You can remove a considerable number of tree squirrels with relatively few traps if you keep them in continuous operation while squirrels are feeding on almonds.

Tree squirrels can also be controlled by shooting, but it is illegal to poison tree squirrels. Consult your local game official for regulations concerning the shooting or hunting of tree squirrels.

Other Vertebrate Pests

Meadow mice. Meadow mice (*Microtus* spp.) also called meadow voles or field mice, inhabit roadsides, canal banks and fencerows (Figure 37). They may move into almond orchards and feed on young tree bark when weeds around the tree trunks offer protection. Meadow mice may also feed just below ground level on the rootstocks commonly used for almonds and can girdle small roots. In addition, they can damage irrigation lines by chewing small holes in them.

Figure 37. The meadow mouse has a blunt nose, small eyes and ears, and a short tail. It is entirely gray or brown and may be up to 6 inches (15 cm) long, not including the tail.

To make a young orchard and its surroundings less favorable to mouse infestation, mow the ground cover or cultivate frequently and control weeds around tree trunks and along fencerows, roadsides and ditchbanks. Although rarely necessary in almond orchards, toxic grain baits are available for meadow mouse control.

Hares and Rabbits. Among the hares, the black-tailed jackrabbit (*Lepus californicus*) is most common as an agricultural pest. True rabbits, the cottontail and brush rabbit (*Sylvilagus* spp.) can, on rare occasions, be pests where trees are newly planted. Both hares and rabbits may feed on the bark of young almond trees, usually only during the winter when other, more preferred, foods are limited. They can also damage irrigation lines by chewing on them.

Treated grain baits are effective in controlling large numbers of jackrabbits or where large areas are involved. Check with your county agricultural commissioner as to which materials may be used in your county. Baits may not be used to control the cottontail and brush rabbits. These species may be controlled by shooting or trapping only. Small populations of jackrabbits may also be controlled effectively by shooting.

In young orchards, individual tree trunk protectors (Figure 38) or fencing is effective against both rabbits and hares. To keep rabbits from slipping through, bury the lower edge of the fence a few inches under the soil and check for holes.

Deer. Deer may damage newly planted almond orchards adjacent to natural habitats, such as wooded and streamside thickets, especially in orchards located in the foothills. They feed at night on young tree foliage and may break small limbs.

Fencing is the most effective method of excluding deer from an orchard. A 6-foot woven mesh wire fence is the minimum recommended height for level ground. A 7-foot fence may be necessary in some areas. If the fence is on sloping terrain or if deer are being chased, they may clear an 8-foot fence. To heighten a 5- or 6-foot-high mesh fence, add two or three strands of wire.

In certain cases, frightening devices may be effective in keeping deer away for short periods of time. Repellants may also prevent deer damage to almond trees. Not all repellants, however, are effective against deer. For more information, contact your farm advisor or county agricultural commissioner.

Coyotes. Coyotes, *Canis latrans*, occasionally cause damage in almond orchards by chewing up irrigation lines. Coyotes are classified as predators and can be taken whenever they are causing a problem. Contact your agricultural commissioner for details about regulations and control options.

Figure 38. Tree trunk protectors will prevent meadow mice and, if high enough, rabbits and hares from feeding on trunks of young trees.

Insects and Mites

Numerous insects and mites inhabit almond orchards. Most cause little or no damage to the tree or crop. Many actually play a beneficial role in the orchard system performing activities such as pollination, feeding on organic debris, or feeding on pests. Only a small fraction of the species present in an orchard cause economic damage to the crop. Of these, navel orangeworm, peach twig borer, San Jose scale, and webspinning (twospotted and Pacific spider) mites have the greatest potential for economic impact on the tree or crop throughout the almond-growing regions in California.

Navel orangeworm overwinters in mature nuts, called mummies, that remain in the tree after harvest. Adult moths emerge in spring and lay first generation eggs on mummy nuts in the orchard. The second generation may also develop on mummies or damaged new crop nuts; but at hull split, egg laying switches from the mummies to the new crop, and the population begins a period of rapid growth. Larvae feed on all parts of the new nuts. Those that reach the kernel devour the nutmeat. Two cultural methods, mummy removal and a rapid, early harvest, can provide effective control of navel orangeworm. In orchards isolated at least 1/4 mile from external sources of infestation, or in localities where area-wide orchard sanitation is practiced, these two procedures can reduce navel orangeworm damage below economic levels and eliminate the need for in-season chemical treatments for this pest. Some damage will occur in orchards where a substantial number of mummies remain from the previous season, regardless of the chemicals used or the area of the state in which the crop is grown. Hence, orchard sanitation, whenever possible, is essential to navel orangeworm management.

Peach twig borer is another major pest of almonds. Larvae overwinter in small cells, called hibernacula, primarily excavated in crotches of 1- to 4-year-old limbs. In spring, they leave their hibernacula, feed on leaf and blossom buds, and mine the tips of branches. Unlike navel orangeworm, twig borer larvae can enter green nuts before hull split. The major period of crop infestation, however, is during and after hull split when twig borer larvae may enter the kernel and feed on nutmeats.

San Jose scale does not directly feed on the crop but damages the tree, eventually causing yield reductions. It feeds on plant juices and contributes to an overall decline in vigor, growth, and productivity. Heavy or severe untreated infestations can kill young trees and twigs and small branches of older trees, reducing vigor and productivity.

The single most effective way to control peach twig borer and San Jose scale is to properly apply a suitable organophosphate insecticide plus superior type oil during tree dormancy. This spray also suppresses European red mite, brown mite, olive scale and European fruit lecanium. Spraying during dormancy causes minimal disruption of natural enemies and should be done yearly. When properly applied, a dormant spray usually eliminates the need for in-season sprays against peach twig borer and San Jose scale. If a dormant spray cannot be applied, alternative control methods, as described in the individual pest sections, may have to be used.

When adequate cultural controls against navel orangeworm are practiced and a dormant spray is applied, in-season sprays for insect control may not be necessary. Some insecticides kill the natural enemies of spider mites and San Jose scale and cause flare-ups of these pests. If you don't spray during the growing season for navel orangeworm, peach twig borer, or San Jose scale, natural enemies will have a better chance of maintaining pest mite populations below damaging levels.

Six species of mites have the potential to cause economic damage in almond orchards. The occurrence of these six mites on almonds and their relative importance varies regionally. Two of these species, twospotted and Pacific spider mites, are considered primary pests and are important throughout the Central Valley. These two species, plus the strawberry spider mite, are collectively referred to as webspinning mites in this manual; all cause a similar type of damage, have similar life histories, are difficult to distinguish in the field, and are controlled in the same manner. Outbreaks are often induced by foliar insecticides; certain orchard conditions also tend to promote mites. Dust from roadsides and frequent cultivation or flailing often contribute to mite outbreaks. If not carefully monitored and treated, webspinning mites can defoliate trees and reduce the following season's growth and yield. Trees with insufficient irrigation will suffer greater defoliation from the same level of mite feeding than trees with adequate irrigation.

In addition to webspinning mites that are frequently a problem in almond orchards, European red, brown, and citrus red mites are occasionally present in large numbers. European red mite occurs throughout the Central Valley. The brown mite is generally found only in Sacramento Valley orchards. Both species can be adequately controlled by a dormant spray consisting of an organophosphate and oil, and usually do not require in-season treatments. Low to moderate numbers of these mites in spring are beneficial because they serve as a food source for mite predators. If mite predators have an adequate food supply early in the season, their population can build up sufficiently to control webspinning mites later in the summer.

Citrus red mite can be a pest in almonds, especially in orchards that are located near citrus groves in the southern San Joaquin Valley. It does not overwinter in almond orchards but re-invades each spring when winds blow it out of citrus groves. Measures taken for webspinning mites in these areas also control citrus red mite, and additional treatment is not required.

Other insect pests of almonds include pavement ant, southern fire ant, oriental fruit moth, peachtree borer, Pacific flatheaded borer, American plum borer, shothole borer, olive scale, European fruit lecanium, boxelder bug, leaffooted plant bug, consperse stink bug, fruittree leafroller, and western tent caterpillar. Many of these insects occur only in localized areas within the almond-growing regions, or they occur infrequently and infestations are not predictable. Of these pests, the ants, oriental fruit moth, and some of the tree borers have the greatest potential for causing economic damage.

Ants are mostly a problem in the San Joaquin Valley, although they can cause damage in all the almond-growing regions. They feed on the nuts after harvest when the nuts are drying on the ground. Ants can sometimes be suppressed by hull split sprays, but severe infestations need to be controlled with a soil-applied granular insecticide.

Oriental fruit moth sometimes injures terminal shoots of trees. Young trees with a limited number of terminal shoots can be severely damaged by shoot dieback. Oriental fruit moth can also be found mining in almond hulls. Although most of their feeding occurs between the hull and shell, larvae occasionally feed on the nutmeats as well.

Peachtree borer, American plum borer, or Pacific flatheaded borer can severely damage or kill a young tree and greatly reduce the vigor of an older one. Peachtree borer and American plum borer can be important as almond pests. Peachtree borer has long been a pest in the Santa Clara Valley and Contra Costa County and is becoming a problem in San Joaquin, Stanislaus, and Merced counties. Larvae enter trees at or near ground level and tunnel in the wood of the trunk. American plum borer occurs sporadically from Merced County to Tehama County. It invades young trees and mines the layer of wood just beneath the bark of the scaffold crotch. As a result, scaffolds frequently break off during windy periods.

Shothole borer infests trees already weakened by some other stress. If the infestation is not severe and the primary source of damage is treated, trees can usually outgrow a shothole borer invasion when they regain vigor.

Two species of scale, olive scale and European fruit lecanium, occur sporadically throughout the almond-growing regions. Neither insect damages trees as seriously as the San Jose scale. Both are controlled by the dormant spray applied to combat the primary pests, and rarely cause any significant damage.

Occasionally an orchard may be invaded in the spring by true bugs that migrate into almonds and feed on the crop until a more suitable host becomes available. Boxelder bugs generally occur in the Sacramento Valley, consperse stink bugs occur in the San Joaquin Valley, and leaffooted plant bugs are more common in the Fresno area. Feeding by these bugs causes small nuts to abort and creates small black depressions on the kernels of older nuts. Many nuts can be affected but these bugs are usually not treated; their occurrence is unpredictable, and by the time damage is noticed they have often left the orchard.

Recently the carob moth, an insect closely related to the navel orangeworm, has been found south of the Tehachapi Mountain Range in California. The carob moth is a pest of soft-shelled almonds, citrus, carob, fig, quince, dates, and pomegranates in the Mediterranean area, but it is too early to predict what impact it will have in California almond orchards.

Fruittree leafrollers and western tent caterpillars feed on the buds and foliage of almond trees, and the leafrollers occasionally get into the green nuts but do not normally cause serious economic damage.

Monitoring Insect and Mites

Monitoring for insects and mites is essential to a good pest management program. Use monitoring results, together with field history and weather data, to identify potential problem areas or "hot spots," to time sprays, to determine the effectiveness of a previous treatment, and to forecast the need for additional control measures. Monitoring programs are available, or are in the process of being developed, for navel orangeworm, peach twig borer, San Jose scale, mites, ants, oriental fruit moth, and some of the tree borers. You must identify pests, their associated damage, and their natural enemies correctly; even closely related species may require different control measures. Photos and descriptions in this manual will assist you in the identifications. For additional help, consult your farm advisor or pest control adviser (PCA).

Monitor for insect and mite pests at the times indicated in Figure 39. Begin monitoring before damage occurs and continue through the pest's damaging stages. Some pests are best monitored during the dormant period. Check every orchard at least once a week and more often during critical periods.

When you monitor insects and mites, keep in mind that different species or different life stages prefer certain parts of the tree or the orchard. For example, brown mites feed on leaves only during the warm period of the day in spring and are found on woody areas of the tree at other times. Infestations of webspinning mites are most likely to begin in dusty, or roadside trees; but European red mite infestations frequently develop in shaded areas with high humidity. Pest development may vary among neighboring orchards due to different cultural practices or the age and cultivar of the trees.

Keep a record of all your monitoring results. Make separate records for each orchard, noting special conditions and pest problems. Record your results on a sampling form, such as the one in Figure 40 (for insects monitored with traps), or Figure 50 (for mites monitored by visual counts).

In addition to counting the actual number of insects trapped, record daily maximum and minimum temperatures and use this information to help you follow insect development with degree-days (see page 16).

Monitoring Methods

The sampling area, or block, will depend upon the size of the orchard, the insect monitored, and the ability to make individual treatment decisions on portions of the orchard. Conditions within each area should be uniform, and all sections of the orchard, except "hot spots," should be included in the sampling scheme. Hot spots, such as the edges of an orchard where dust or insecticide drift may lead to a buildup of mites, should be monitored and treated separately. At each sampling area, look for natural enemies of mites as well as for pest species. Sufficiently high populations of natural enemies can rapidly bring a potentially damaging mite population under control.

As a general guideline, divide large orchards into 80-acre sampling blocks. Orchards that exceed 1,000 acres can be divided into larger sampling blocks if conditions within each block are uniform. Within each block, choose four sampling areas, and place two navel orangeworm egg traps, one peach twig borer pheromone trap, and one San Jose scale pheromone trap in each area. If you are monitoring for oriental fruit moth as well, place one oriental fruit moth trap in each area. For convenience, hang the peach twig borer, San Jose scale, and oriental fruit moth traps on the same tree with one of the navel orangeworm traps. Place the second navel orangeworm trap about 10 trees away from the first. In orchards that are 80 acres or less, select four sampling areas, and place one navel orangeworm, peach twig borer, and San

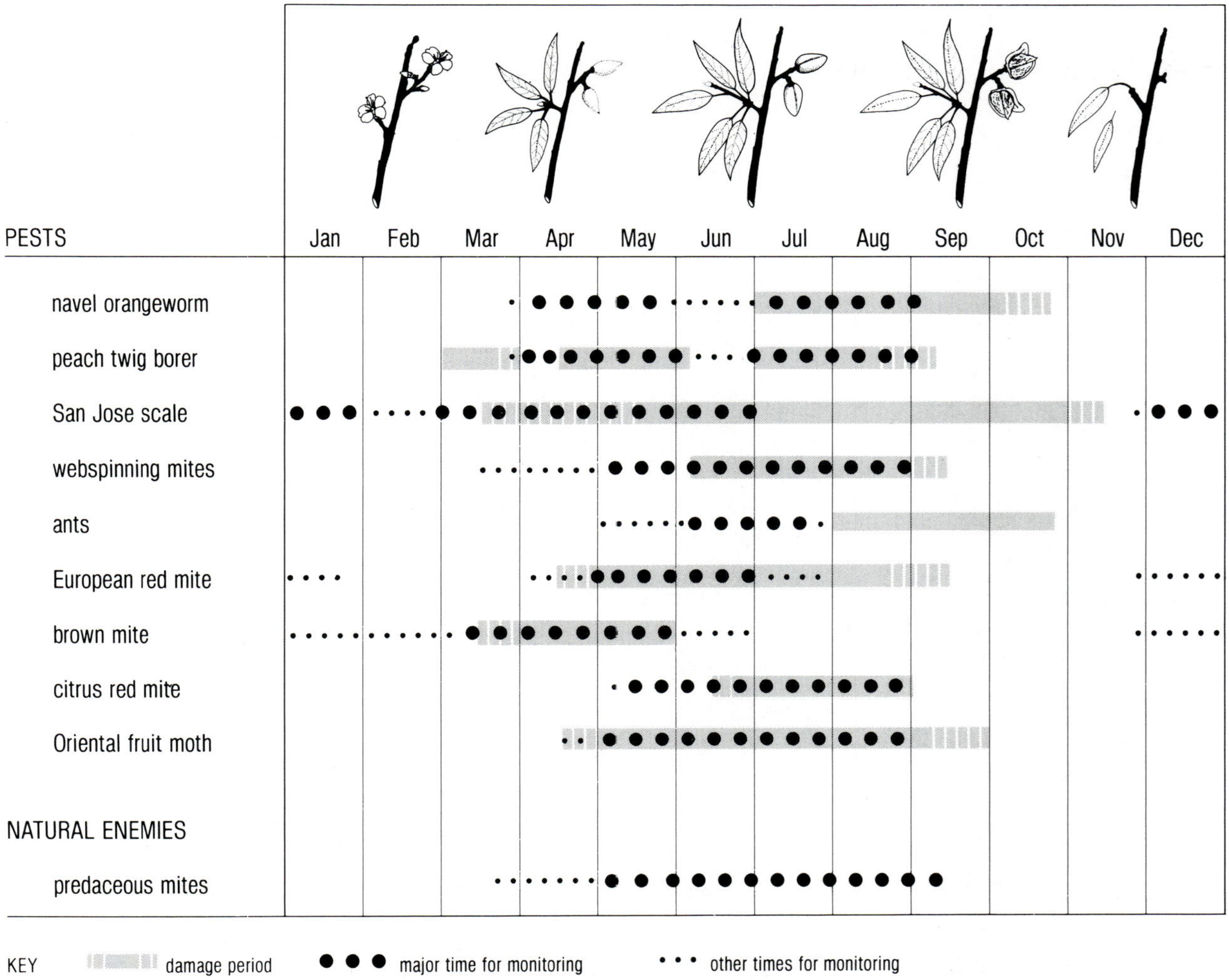

Figure 39. Monitor insects and mite pests when they are likely to cause damage or at other times indicated. Developmental times for the almond branch across the top of the figure are for an early cultivar; events may occur up to one month later in late cultivars.

Jose scale trap in each area. If you are also monitoring for oriental fruit moth, use one trap at each location for this moth as well. You can use trees with traps and adjacent trees to monitor other pests, such as mites and tree borers.

These guidelines represent the ideal. The actual number of traps used in a particular orchard will also be influenced by other criteria, such as time available to check them and whether or not the sampler is monitoring several orchards in the same vicinity, thus reducing the number of egg traps needed in each orchard.

Visual Sampling. Some pests are monitored by thoroughly examining leaves, trunks, twigs, or nuts for eggs, larvae, adults, or damage. Check for scale insects in the tops of trees during the dormant season. To detect tree borer infestations, examine the tree trunk and scaffold crotch carefully for entry holes or frass. The type of borer being monitored will determine the location on the trunk to look for evidence of an infestation. Examine the orchard floor for ants.

Assess mite populations by sampling leaves to determine the presence or absence of mites and mite predators or brushing and counting the number of mites per leaf (see the monitoring section for mites in this chapter). Mites may first appear in hot spots, such as dusty border areas of the orchard. Do not base a treatment decision for the entire orchard on samples taken from these areas. Spot-treating localized infestations will preserve natural enemies in other areas of the orchard.

Although egg traps are used to follow seasonal development of navel orangeworm (see below), the sever-

INSECT MONITORED ______________________________

ORCHARD LOCATION ______________________ **BOTTOMS CHANGED** ______________________

TRAPPING PERIOD ____________ to ____________ **CAPS OR BAIT CHANGED** ______________________

COMMENTS ______________________________

	Temperature (°F)		Degree-days (°D)		INDIVIDUAL TRAP COUNTS					AVERAGE TRAP COUNTS
Date	Max.	Min.	For day	Accum.	Trap 1	Trap 2	Trap 3	Trap 4	Trap 5	Pests/trap/day

Figure 40. Record data from traps on a record sheet, such as this one. Use a separate form for each insect—that is, navel orangeworm, peach twig borer, San Jose scale, and oriental fruit moth.

ity of infestation by navel orangeworm is correlated with the presence of mummies on soft-shelled cultivars. These are easily seen in winter.

Traps. Monitoring programs for navel orangeworm, peach twig borer, San Jose scale, and oriental fruit moth use traps that attract insects by scent. These traps can help you detect the presence and seasonal development of these pests in the orchard. You can use trapping information to time pest control actions more effectively. The number of traps needed depends on the size of the orchard and the insect being monitored. For convenience, place the traps for the four species in the same or nearby trees at a sampling site.

The overwintering generation of navel orangeworm normally lays eggs on mummy nuts. Egg-laying activities can easily be monitored with egg traps; these are small plastic containers with two large windows that are covered with nylon organdy screening. The bait is pressed almond cake. It attracts female moths, who lay eggs around the rim of the trap cap, on the grooves of glue that secure the organdy screening, on roughened areas around the windows, and on the bottom of the trap. Traps painted with flat black spray paint are more effective than white traps.

Egg traps can help you determine when the navel orangeworm is first active in the spring so that a spring spray, if needed, can be timed properly. Traps cannot help you determine if a spray is necessary: the number of eggs found on traps has not been correlated to the subsequent magnitude of navel orangeworm damage.

Pheromone traps are used to monitor peach twig borer, San Jose scale, and oriental fruit moth activity. Pheromones are attractants produced by animals to communicate with individuals of their own species. Pheromones currently being used in IPM programs are sex pheromones released by females to attract the males for mating.

Most commercially available pheromone traps are made of folded cardboard that is coated on the lower inside with a sticky substance. Inside the trap is a replaceable plastic or rubber cap impregnated with a pheromone. The scent is slowly released from the cap and attracted insects get stuck on the sticky surface of the cardboard. Sex lures are fairly specific, although other species are sometimes caught. Pheromone traps require regular changing of the caps and sticky liners, but they are easier to service than most other trap types, such as blacklight traps or liquid bait traps, and are more economical.

Flight activity varies with environmental conditions, and trap catches may vary with weather. For example, cool temperatures or rainy weather may result in low trap catches. Be sure to correlate trap catches with weather information.

Prevention and Management

Cultural Practices. Orchard sanitation is an important management tool for reducing the buildup and spread of insects. Removing mummy nuts from the trees and destroying them deprives the navel orangeworm of overwintering sites and egg-laying sites in the spring. Also, removing fruit during the winter from nearby alternate hosts such as loquat, prune, and peach can reduce navel orangeworm populations by eliminating these potential overwintering sites. Harvesting early reduces navel orangeworm infestations because navel orangeworm moths will not lay eggs on nuts that are on the ground. Adequate irrigation enables mite-infested trees to withstand feeding by greater numbers of mites before economic loss occurs than mite-infested trees that are not receiving sufficient irrigation. Finally, pruning trees to remove damaged or dead limbs will reduce invasions by wood-boring insects.

Biological Control. Potential pest populations in almond orchards may be controlled by other organisms inhabiting the orchard. These biological control agents include predators, parasites, and pathogenic microorganisms. Numerous native predators and parasites of almond pests occur throughout the Central Valley. In addition, exotic natural enemies of the navel orangeworm (those found in its natural home range) have been collected and introduced into the Central Valley by University of California scientists.

Natural enemies are very important in the control of mites and scales but do not provide economic control of other major almond pests. Natural control of scales and mites is easily disrupted by in-season chemical treatments, resulting in outbreaks. You can reduce chemically induced mite outbreaks by carefully selecting types and rates of insecticides and timing applications so that they are least disruptive to natural enemies.

Pesticide Sprays. Correct application of a needed pesticide is essential to a successful integrated pest management program. To achieve effective pest control, you must choose the correct coverage, ground or air speed, gallonage, and type of equipment (see the section on pesticides in the chapter, "Managing Pests in Almonds.") For information on the type and rate of pesticide to use, see *A Guide to Controlling Almond Pests, Diseases, and Micronutrient Deficiencies,* listed under Pesticides in References.

Dormant Sprays. As a rule, all orchards should receive a dormant spray each year to control peach twig borer, San Jose and other scales, and mites that overwinter in the orchards as eggs. Although a dormant spray applied just

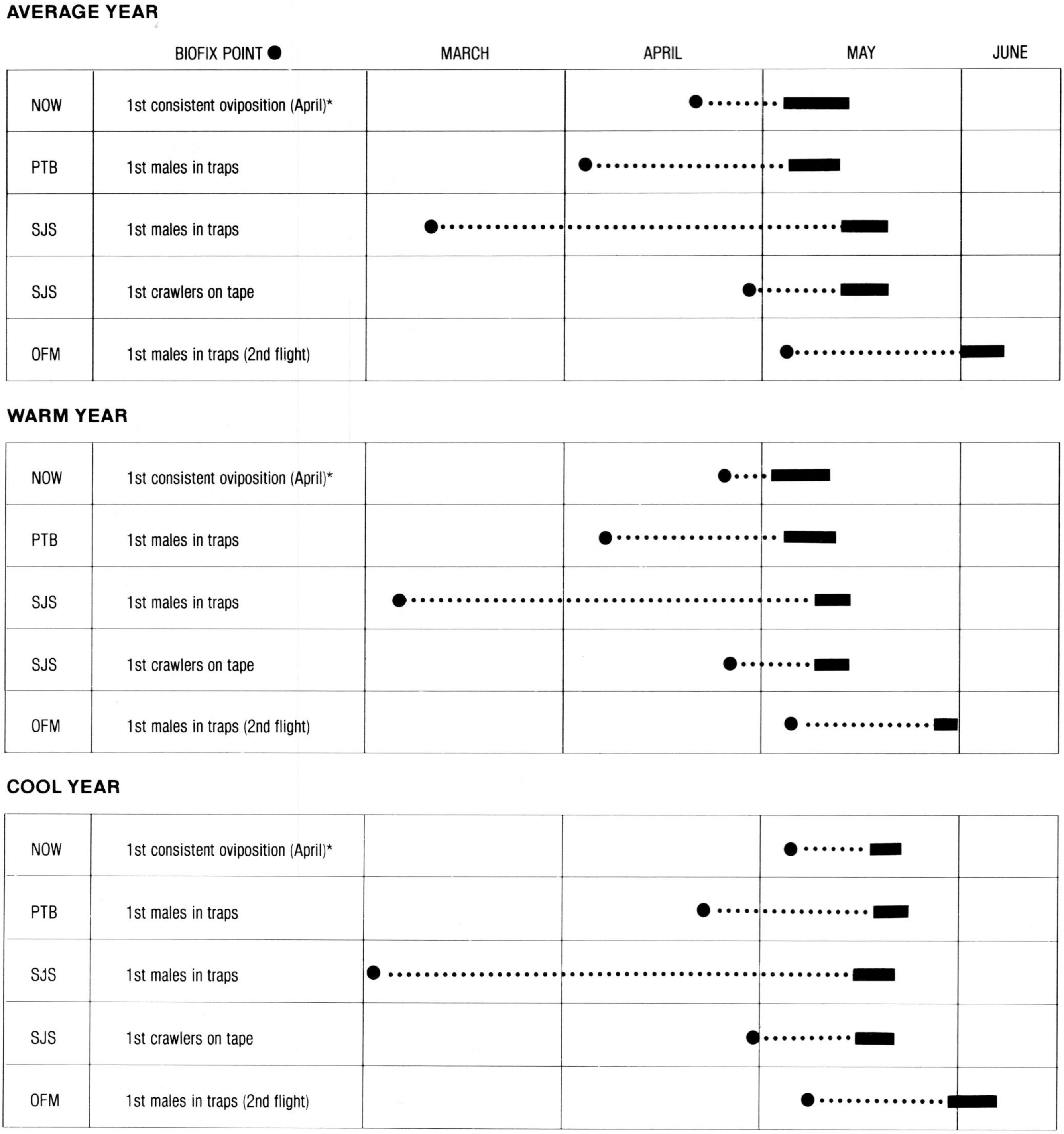

*Egglaying for 1st navel orangeworm brood typically begins after April 10.

Figure 41. Optimum spray timings for navel orangeworm (NOW), peach twig borer (PTB), San Jose scale (SJS), and oriental fruit moth (OFM) for three separate years in Fresno County: an average year, a year in which temperatures were warmer than average, and a year in which temperatures were cooler than average. Note the synchrony of the four species to each other and the variation in timing from one season to the next. San Jose scale has two biofix points, depending on the trapping method used. The bar following each biofix point represents the optimum spray timing for that insect.

before or at bud swell may be most effective, early spring rains may prevent access to the orchard. Because a dormant spray is basic to a good pest management program, it is more important to get it on than it is to apply it at the delayed dormant stage. It can be applied any time after the first heavy rain and the leaves fall in the winter. A heavy rain will ensure that the trees are not moisture stressed, thus avoiding possible phytotoxic effects of the oil on the trees. Absence of leaves will allow best coverage with the spray.

Use an oil plus one of several organophosphate insecticides. You can use superior, supreme, or one of the dormant type oils in your spray. Oil helps the insecticide penetrate to target areas, and it also is more effective than the synthetic insecticide in killing mite eggs.

In-Season Sprays. Use the guidelines in this chapter to determine which in-season sprays, if any, you will need to apply. The optimum spray timing for a particular pest may vary from year to year; it can also vary in regard to the optimum spray timing for other pests. Figure 41 illustrates examples of optimum early season spray timings for major almond pests, based on degree-day accumulation techniques outlined in this chapter, for three different years. To effectively time pesticide applications, always monitor the pests you need to control.

Navel Orangeworm
Amyelois transitella

Navel orangeworm is the primary insect pest of almonds in California. A native of the southwestern United States and Mexico, it was first discovered feeding on damaged navel oranges in Arizona during the early part of this century. Although this was the origin of its name, navel orangeworm is not an economic pest of oranges. It feeds on the dried, decaying, or damaged fruits of many trees including citrus, stone fruits, figs, and pomegranates, as well as a number of noncrop plants such as acacia and locust. In almonds, walnuts, and pistachios, however, navel orangeworm attacks healthy nuts after hull split and is a major pest.

Description

Adult moths have irregular silver gray and black patterns on the forewings and legs. The navel orangeworm belongs to the pyralid, or snout moth family, and has a small, snoutlike projection formed by a pair of palps in front of the head. In cool, wet spring weather adults may live 2 to 3 weeks; as the season progresses and days become hot and dry, they survive less than a week. Most female moths mate within 2 days of emergence and lay an average of 85 eggs per female over a period of 2 to 8 nights, depending upon the temperature. No eggs are laid when temperatures drop below 55°F.

First generation and some second and third generation eggs are laid singly on the surface of mummy nuts; occasionally they are also found close to the nut on twigs or nut stems. Most of the second, third, and fourth generation eggs are laid on the new crop nuts. When first laid, the eggs are creamy white, flattened, and oval with irregular reticulations on the surface. The color changes from white to pink and then to reddish orange just before hatching. At this time, the black head capsule can also be seen inside the egg. Eggs hatch in 4 to 23 days, depending on temperature. Development of all life stages is temperature dependent and occurs more rapidly during warmer weather.

Newly-hatched larvae are the same reddish orange color as the mature egg. After feeding begins, larval color is influenced by the nature of food and can range from milky white to pink. The head capsule is dark reddish brown in all instars. A pair of crescent-shaped marks on the second segment behind the head distinguishes navel orangeworm from most other larvae, except carob moth (see page 90). Upon reaching maturity, larvae spin a closely woven cocoon and pupate.

Navel orangeworm pupae are light to dark brown and are found inside nuts or between hulls and shells, where the larvae have been feeding. The pupal stage lasts from 6 to 10 days in hot weather to several weeks in late winter or early spring.

Seasonal Development

Moths of a given generation may be active and lay eggs over a period of about 2 months, depending on weather. These activity periods are called flights. The navel orangeworm has three and sometimes four flights each season. The fourth flight occurs in years when temperatures are warmer than usual and degree-day units are accumulated more rapidly than in cooler years. Figure 42 shows the typical seasonal development of the navel orangeworm.

Navel orangeworms overwinter as larvae in almond mummies that remain on the trees after harvest. Larvae do not usually enter diapause, so feeding and development may occur during the warmer periods of winter. Larvae pupate within the mummies in early spring, and moths of the overwintering brood begin emerging in late March or April, marking the beginning of the first flight. Emergence and egg laying by the overwintering generation usually peak in May and are completed by early to mid-June. First generation eggs are laid on mummy nuts remaining in the trees, and larvae develop exclusively in these nuts.

First generation female moths that emerge in mid- to late June for the second flight lay their eggs on mummy

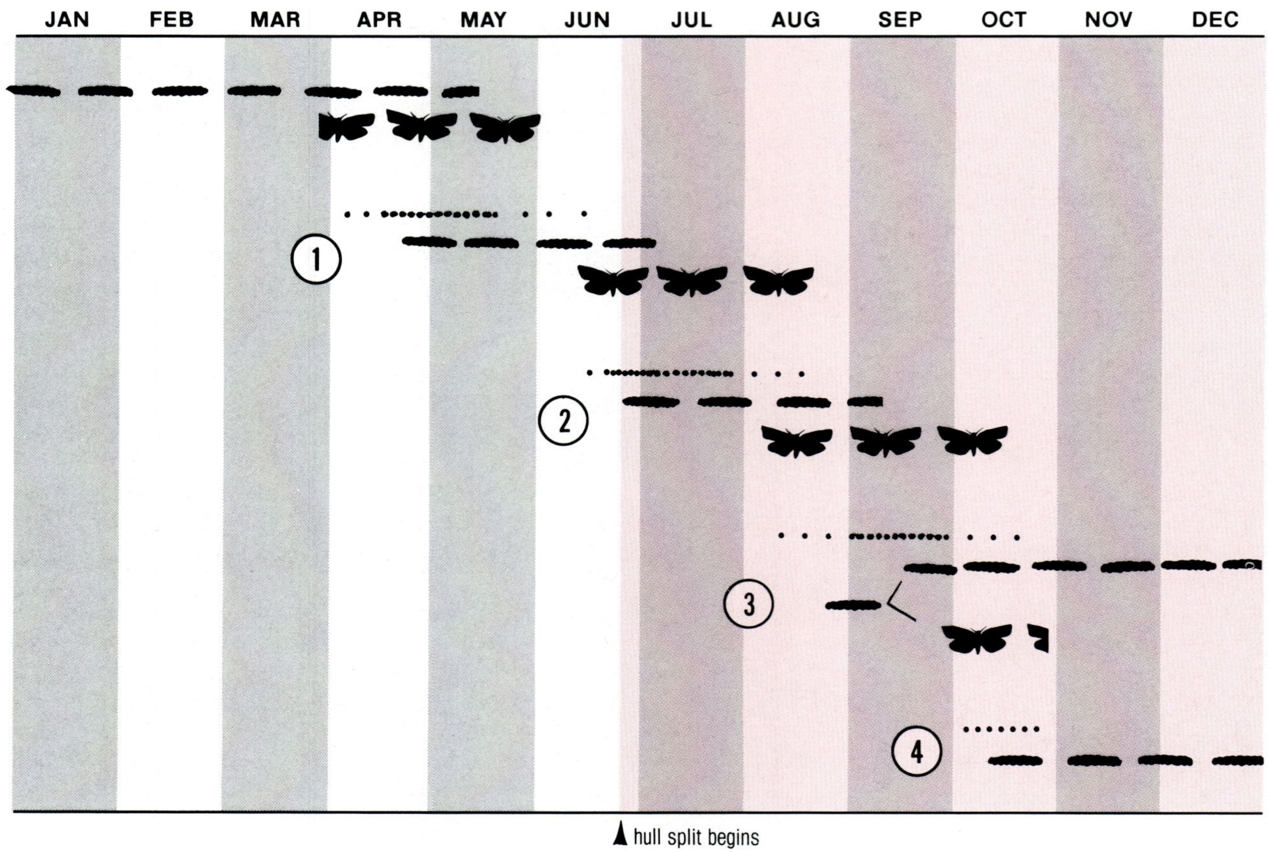

Figure 42. Seasonal development of the navel orangeworm. Egg laying by navel orangeworm varies from year to year and from district to district. For the egg stage, the denser dots indicate the period during which egg laying typically peaks. The date hull split begins can also vary from one year to the next, depending upon temperature. The fourth flight and the partial fourth generation are more common in southern portions of the Central Valley, especially in years with above average temperatures.

nuts or occasionally on green nuts damaged by peach twig borer or oriental fruit moth. Navel orangeworm larvae can then enter these damaged nuts before hull split and go directly into the kernel to feed.

At the onset of hull split in the Nonpareil cultivar, egg laying begins on the new crop nuts. As the infestation develops, moths prefer to oviposit on nuts previously infested by navel orangeworm or peach twig borer, although they readily attack uninfested nuts. In soft-shell cultivars larvae penetrate the shell and feed on the nutmeats, but they are unable to penetrate the shell of hard-shell cultivars. Larvae can also feed and complete their development in the hull. Although a single larva in an almond of the new crop is most common, often there are two or three, and as many as 17 have been found in a single nut.

Second generation larvae that have developed on the Nonpareil crop during July and August may emerge as moths (third flight) in time to lay eggs on and reinfest this crop—hence the importance of a timely harvest. They can also infest later maturing crops of pollinizer cultivars, such as Merced or Thompson, still to be harvested. Navel orangeworm populations can grow rapidly through the remainder of the season because the new crop nuts are a much better and more plentiful food supply than mummies from the previous year, and because temperatures are usually warmer during July and August than in the late spring and early summer. As a result, the second generation of moths (third flight) is significantly larger than the overwintering or first generation, and they lay eggs for the third generation, which has the potential to cause the most nut damage.

In unusually warm years, some of the early third generation eggs may be able to complete development and give rise to a small moth flight. This mostly occurs in the central and southern portions of the San Joaquin Valley in years of above average temperatures. These moths lay eggs for a fourth generation. Eggs hatch and larvae overwinter along with some of the older third generation larvae.

Adult navel orangeworms are dull, grayish brown moths with irregular silver gray and black patterns on the forewings and legs. A pair of palps in front of the head form a snoutlike projection.

Older navel orangeworm larvae have a dark reddish brown head capsule and range in color from milky white to pink.

Damage

Navel orangeworm larvae damage almonds by feeding directly on the nutmeats. First instar larvae that bore directly into the nutmeat leave a minute quantity of frass at the point of entry, which is called a pinhole. Larvae that feed first on the hull or outer parts of the shell and bore into the nutmeat in the second or later instar do not produce a pinhole. As the larvae feed and get larger, they consume most of the nut and produce large amounts of frass and webbing.

Nuts infested in July are often completely consumed by harvest. Most of the completely consumed nuts are blown out by the pick-up machines and hullers and are not recorded as worm damaged in harvest samples. Eggs laid on nuts just before harvest will hatch, and the larvae will develop in harvested nuts that are not fumigated immediately. Besides damaging nutmeats and leaving frass and webbing, navel orangeworms cause nuts to be

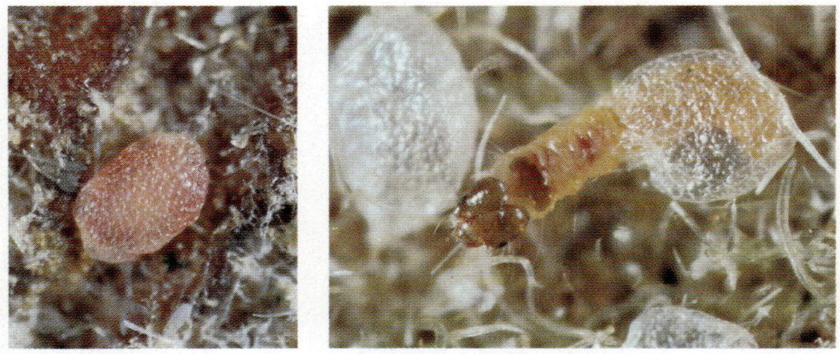

Navel orangeworm eggs are creamy white when first laid on the almond hull. The color changes from white to pink and then turns reddish orange just before hatching (left). Newly hatched navel orangeworm larvae are the same reddish orange color as the mature egg (right).

The navel orangeworm larva has a pair of crescent-shaped marks on the second segment behind the head; the presence of these marks distinguishes it from other larvae such as the oriental fruit moth, but not the closely related carob moth.

Navel orangeworm larvae consume the almond and produce large amounts of frass and webbing. A dark brown pupae can be seen encased in the webbing and frass in the center of the photo.

NAVEL ORANGEWORM

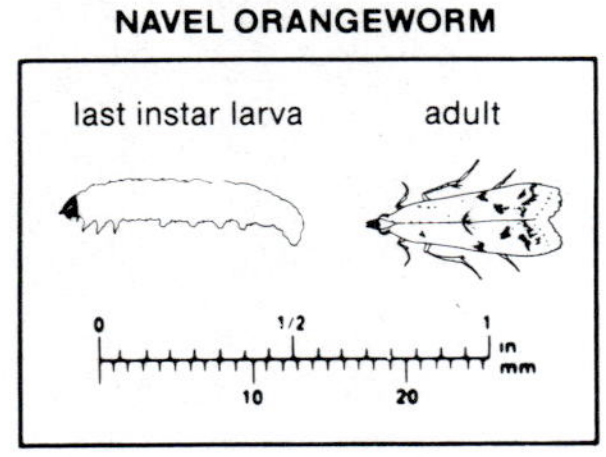

Mummies, or nuts that remain on the tree after harvest, provide overwintering sites for navel orangeworm larvae.

One method of removing mummy nuts is to shake the trees during winter.

more susceptible to fungal infections that can lead to aflatoxin production, a serious threat to the marketability of any crop.

Some cultivars of almonds are more susceptible to navel orangeworm damage than others. Shell hardness and seal and time and duration of hull split all influence susceptibility. Soft-shell cultivars with a poor seal are more susceptible to navel orangeworm infestations than hard-shell cultivars with tight seals. Later maturing, soft-shell cultivars with a lengthy hull split period are also very susceptible to attack because the nuts are still on the trees when third generation moths emerge in August and September to lay eggs. Therefore, late maturing cultivars with a lengthy hull split period, soft-shells, and poor seals show the highest levels of navel orangeworm damage. In recent years industry-wide nut reject levels, which are primarily due to navel orangeworm damage, have been highest in Jordanolo, Merced, and Thompson cultivars, followed by Nonpareil, Ne Plus Ultra and Fritz, in that order. Ruby, Peerless, Butte, and Price are relatively resistant, while Mission and Carmel are the most resistant to navel orangeworm.

Shell hardness, seal, length of hull split, and therefore, susceptibility of cultivars may vary somewhat from year to year, depending on growing conditions. Also, navel orangeworm infestations may vary according to crop set. A light crop may have a heavier infestation than a heavy crop because navel orangeworm activity is concentrated on fewer nuts. The result may be a higher percentage of damaged nuts in a given year, even though the size of the navel orangeworm population is the same as in previous years.

Management Guidelines

Orchard Sanitation. Navel orangeworm overwinters and spends the first generation in spring in mummy nuts. The most effective way to prevent economically destructive populations of navel orangeworm is to remove mummy nuts from the trees by February and destroy them. Mummies on the ground should be destroyed by March 15. When a good orchard sanitation program is carried out in an orchard at least 1/4 mile from infested trees, together with an early harvest, usually no sprays are needed for navel orangeworm. Even when chemical control programs are planned, partial mummy removal will reduce navel orangeworm damage.

Remove mummy nuts from the tree either as part of normal harvest operations or during the winter (December and January). If mummies are removed as part of harvest, returns from recovered nuts may help offset the cost of cleaning.

Mummy nuts can be removed from the tree by mechanically shaking the tree or by hand poling. If an average of more than two mummies remain on trees of soft-shelled cultivars in January, pole or shake again. Generally, trunk shaking is cheaper than hand poling and is better for trees over 12 feet tall, if the orchard soil is not too wet for the shakers. Loss of a few fruit buds from mechanical shaking does not significantly affect the following season's yield. Hand poling may be efficient for trees less than 20 feet tall and with fewer than 50 mummies per tree, but it should be done before bud swell or it may remove a significant number of buds. Bud removal may slightly reduce crop potential—especially on Merced and Thompson cultivars, where many of the mummy fruits and buds are located along main limbs.

Remove mummies in the winter under the wettest possible conditions, preferably after 1 or 2 days of wet fog or drizzle. Moisture softens the fibers holding the mummies on the tree and also soaks into the mummies making them heavier and easier to remove. Fog and moisture during the winter months may be limited, making cleanup hard to complete. When optimum conditions occur, first knock the cultivars that are most susceptible to navel orangeworm infestation and most difficult to clean—such as Merced and Thompson. These cultivars are prime overwintering sites for navel orangeworm. Less susceptible cultivars can be knocked later. Some growers conducting winter cleanup take maximum advantage of the number of hours of fog or heavy dew by shaking trees at night or in the early morning.

Other cultivars such as Nonpareil, Ne Plus Ultra, and Price are easier to clean. Mummies on Mission trees should also be removed because navel orangeworm larvae can overwinter between the hull and the shell of these nuts. Mummies on Peerless trees generally fall off the trees and cleaning this cultivar is not essential, but it will help to concentrate bird activity on other mummies in the orchard that might harbor navel orangeworm.

Birds feed on mummy nuts and often remove substantial numbers during the winter. If you clean the orchard in early December, the birds will concentrate their activities on any remaining mummy nuts over a longer period of time. Birds may be significant in the final stages of orchard cleanup in orchards near river bottoms, in flyways, or adjacent to other crops where they often rest or feed. In orchards that are in areas of large, solid blocks of almond plantings, however, bird activity may be inconsequential.

Mummy nuts remaining on the ground after harvest or knocking should be destroyed by the time moths start emerging in mid-March. You can still benefit from mummy destruction as late as May 1, but for best results, mummies should be destroyed by March 15. Discing or flail mowing between rows, a normal operation in many orchards, along with wet conditions that are conducive to rotting, will kill most navel orangeworms present. On bare ground where herbicides are applied over the entire

Mummy nuts can also be removed by knocking them from the trees with poles.

Monitor navel orangeworm egg-laying activities with an egg trap baited with almond press cake.

orchard floor or on herbicide-treated strips, berms or permanent irrigation checks, blow or sweep nuts to the middle of the row before discing or flailing.

Unkept orchards and backyard fruit and nut trees are a source of reinfestation when they are near a commercial orchard and should be cleaned up if possible. Sanitation is most effective when it is carried out on an area-wide basis. For maximum effectiveness from a sanitation program, an orchard should be at least 1/4 mile from external sources of infestation. Even in nonisolated orchards, however, sanitation is beneficial and economical.

Rapid, Early Harvest. Getting Nonpareil and other soft-shelled almonds off trees, picked up, hulled, and to the handler for fumigation as early and quickly as possible is essential for effective navel orangeworm control. Even in orchards that are cleaned and that have received navel orangeworm sprays, damage can be high if early maturing, soft-shelled almonds are not harvested as early as possible. This is because egg laying by the third navel orangeworm flight coincides with the optimum time of harvest.

For an early harvest, you must take into account not only the *degree of hull split* and the *percent of nut removal,* but the start of the *third navel orangeworm flight* as well. The earlier you harvest your soft-shelled crop, the less time it will be exposed to the damaging third generation larvae, and the lower your infestation will be. When 95 to 100% of the almonds at eye level are visibly hull split (Stage 3 or more, see Figure 6, page 12), shake a few trees to determine if nut removal is satisfactory. If too many nuts remain in the tree, try again in a few days. When at least 95% of the nuts fall when trees are shaken, begin harvest. Hulls will still be green but will dry rapidly on the ground. Navel orangeworm moths do not lay eggs on nuts on the ground.

If early maturing, soft-shelled almonds are removed from the orchard in adequate time, much of the navel orangeworm population will be removed with them, thus reducing the third moth flight. Early harvest of later maturing, soft-shelled cultivars also prevents overlap of generations on the same crop.

If you can't harvest an entire orchard as early as desired, first harvest those blocks that had a higher mummy load at the beginning of summer or are near external sources of navel orangeworm. Harvest susceptible, soft-shelled cultivars, such as Nonpareil, Thompson, Merced, and Ne Plus Ultra, as early as possible.

If possible, do not delay Nonpareil harvest beyond the beginning of the third flight; a second shake can be performed at less cost than the additional insecticide treatment that is needed if harvest is delayed. To detect the beginning of this flight, use egg traps and inspect green nuts for the first new eggs, which will indicate that the third flight has begun.

Almonds can be picked up after harvest as soon as the hulls are dry enough for hulling. The length of time for almonds to dry on the orchard floor depends on a variety of factors such as the degree of shading, the cultivar of the tree, how soon before harvest the trees were irrigated, and how soon after hull split the nuts were harvested.

A good rule of thumb to determine hullability and field storage is to hold the hull where it is split from the shell between the thumb and index finger and bend it back on both sides. If it snaps, it will hull properly. If the hull just bends back, the nuts are not ready for the huller. Check 10 randomly selected nuts from several locations in the orchard. If at least 80% of them snap, the nuts can be delivered to the huller.

If you are unable to harvest before the beginning of the third flight, pick up the almonds as soon as possible after knocking and fumigate them. Fumigating them at this time destroys larvae while they are small and before they damage the kernel. Otherwise, navel orangeworms already present in these almonds will continue to develop and cause damage. More importantly, adults can emerge from these nuts to infest later maturing pollinizer crops still in the trees. Rapid removal of nuts at harvest also will help reduce ant and squirrel damage.

If possible, fumigate or hull worm-susceptible cultivars first. Until the hulls and meats are separated, eggs and small larvae on the hulls remain a threat to the nutmeats.

On-Farm Fumigation. Consider on-farm fumigation immediately after harvest. Although almonds are fumigated as a standard practice upon receipt by handlers, the interval between harvest and handler fumigation can be several weeks. During this time, navel orangeworm and peach twig borer larvae already present in the nuts can continue to develop and cause more damage. Even if the crop is sent directly to the huller, enough time may elapse that on-farm fumigation will provide significantly less insect damage. On-farm fumigation is simple, inexpensive and will halt activity of navel orangeworm and other insects. Follow the guidelines for fumigation in *Fumigation of Inhull Almonds on the Farm,* listed under Pesticide References.

Biological Control. Several parasitic wasps are currently being studied for their effectiveness in controlling navel orangeworm populations.

Pentalitomastix plethorica was introduced into California from Mexico in the 1960s and is now widespread in California. *Goniozus legneri* was imported from Uruguay and Argentina and released in California orchards in 1979. Both wasps coexist in the almond orchards and together appear to cause a greater mortality of navel orangeworm than either one alone. It is too early to evaluate their effect on navel orangeworm population levels.

Although birds are generally considered orchard pests in the summer, they do remove mummy nuts from the trees following harvest and are considered beneficial during the winter months. Their effectiveness is determined by the orchard's proximity to areas where birds normally congregate in the winter months.

Monitoring and Control. In almond orchards where an adequate sanitation program has not been carried out, or where there is a nearby external source of navel orangeworm, an insecticide application may be necessary during the spring period of egg laying in May, and at initiation of hull split. If you plan only one treatment, spray at hull split only. The hull split spray is easier to time than the spring spray, and it does not disrupt natural control of mite pests in the early season. If you also need to control peach twig borer, however, apply your spray in spring and be sure to monitor mite populations afterwards. A miticide treatment may be necessary if mite predators are absent and your orchard has a history of mite problems or an early season spray was applied to control other pests. A miticide treatment can be combined with the hull split spray if you plan to spray at that time. Orchards not harvested early may need an additional spray before harvest. This additional spray, however, is not as cost effective as early harvest and orchard sanitation. Also, harvest-interval requirements must be strictly followed.

Manage your orchards to avoid spraying for navel orangeworm; sprays can cause serious outbreaks of mites and destroy natural enemies of the navel orangeworm and other insect pests. In years with a good crop set, an orchard that meets all the criteria listed below will probably not need an in-season spray to deliver a crop with a tolerable level of orangeworm damage (up to 4%) on Nonpareils. (In years when the crop set is light, however, damage may be higher because feeding by the navel orangeworm population will be concentrated on fewer nuts.) Sprays may not be needed in an orchard that:

1. Has soft-shelled cultivars with less than an average of two mummies per tree on February 1. Count the number of mummies on two randomly selected soft-shell trees per acre to determine the average number of mummies per tree. Check a minimum of 15 trees per block.

2. Has mummies on the ground destroyed by March 15.

3. Is isolated by at least 1/4 mile from external sources of infestation.

4. Will be harvested promptly—knocking just as the crop reaches 100% hull split, or at least before second generation (third flight) moths emerge.

Spring Sprays. The best time to apply a spring spray is just after the first eggs of the spring brood hatch. The beginning of this brood will vary according to year and location. Observe eggs deposited on traps to determine when egg hatch begins. However, egg trap counts will not tell you if a spray is needed. Evaluate your orchard using the criteria listed above to determine if you need to spray.

Almond press cake is usually used as the egg trap bait to attract female navel orangeworms to lay eggs on the traps. If kept dry and powdery, the press cake can be used up to 8 weeks without change; replace it when it gets wet or lumpy and moldy.

Place egg traps in orchards by the first week of April. Use at least four traps per orchard, and in large orchards use an average of one trap per 10 acres. Divide large orchards into 80-acre sampling blocks. Each block should be subdivided into four sampling areas and two traps

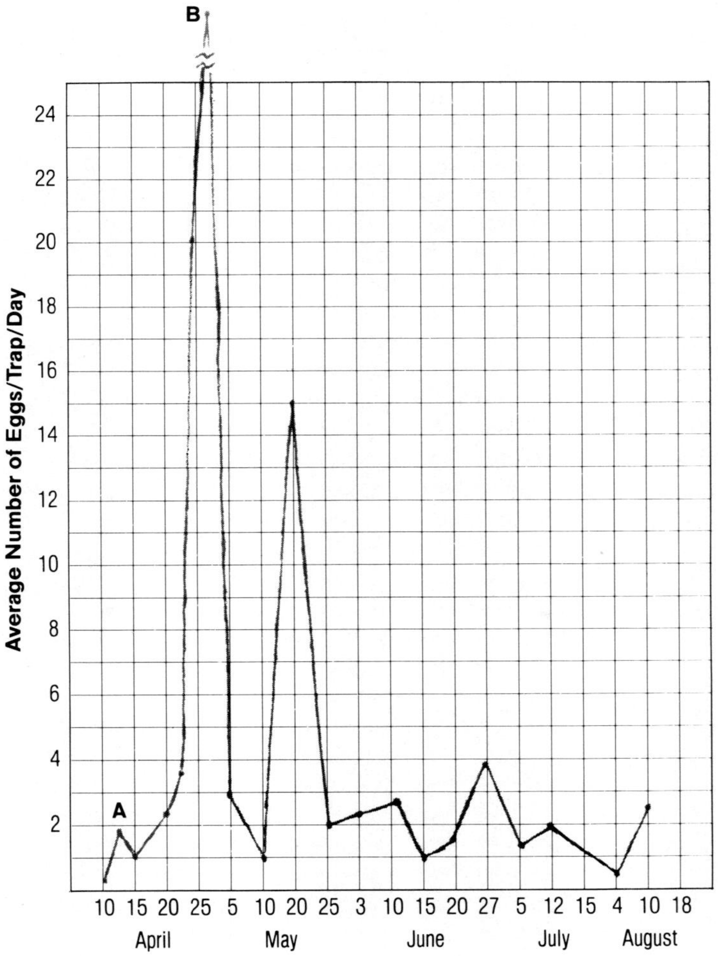

Figure 43. Graph of navel orangeworm egg trap catches. Egg laying for navel orangeworm brood typically begins after April 10 and is characterized by a definite increasing trend in egg laying (B). This increase is generally preceded by a slight increase in egg laying (A).

hung in different trees in each area (see page 50). Orchards that exceed 1,000 acres can be divided into larger sampling blocks if conditions within each block are uniform.

Hang egg traps at head height on the north side of Nonpareil trees, 1 to 3 feet inside the drip line. Do not place traps in border trees. In sprinkled orchards, install low-angle heads in trap areas, or hang the traps over the sprinkler head to keep the press cake bait dry.

Preceding the major spring flight, a low level of sporadic activity may occur between late March and mid-April. Check the traps twice weekly and graph the number of eggs per trap per night, as shown in Figure 43, to visualize orangeworm activity. The actual start of major spring activity is characterized as a consistent, increasing trend in egg laying, which typically begins after April 10 and coincides with warming temperatures. Watch the graph and remove any eggs laid on the traps before a definite increase in egg laying. (A toothbrush works well for this purpose.) If egg laying appears to be increasing, and the next trap check in 3 or 4 nights confirms that it is, circle freshly laid white eggs on two or three of the traps with a wax pencil, remove the bait from each trap, hang them in varied locations in the tree canopy, and watch the eggs to see when they hatch. With typical spring weather, eggs will hatch about 10 days after they were laid. If the weather turns cold, hatching will take longer. The correct time to apply a navel orangeworm spray is when marked eggs begin to hatch.

Meanwhile, to confirm that a definite spring brood has begun, continue to record egg laying from the rest of the traps, and graph egg trap data until the peak period of egg laying is clearly under way.

If your spray emphasis is both navel orangeworm and peach twig borer, use monitoring information for navel orangeworm to time sprays in May. Peach twig borer populations have been known to peak a few days *after* the navel orangeworm spray should have been applied. Spring sprays timed for navel orangeworm based on egg trap counts generally give good twig borer control; separate May sprays for the two insects should not be needed.

Table 7. Effects of Different Insect Management Practices on Reduction of Crop Rejects Caused by Navel Orangeworm (From a 1981 survey of California almond growers).

Insect management practice[1]	Estimated navel orangeworm reduction in rejects (percent)[2]
1. Winter cleanup of mummies (2 or less mummies/tree)	
a. orchards near external source of infestation	50–70
b. isolated orchards	70–80
2. Early harvest (100% hull split)	70
3. la and 2 combined	75
4. 1b and 2 combined	80–90
5. Properly timed May spray (miticide sometimes included)	40–60
6. Properly timed hull split spray (miticide included)	40–60

1. It was assumed that a dormant spray of an organophosphate insecticide and a narrow range oil was applied for peach twig borer.
2. Use these percentages as guidelines only. Results obtained by growers may differ slightly. Adjust the control percentages to reflect the results typically obtained in your own orchard with each of the insect management practices listed.

Hull Split Sprays. Unlike the spring spray for navel orangeworm, which is timed to egg hatching, the summer application is timed to nut development. The best time to spray for navel orangeworm is when hull split begins on sound nuts in the *tops* of the trees. The nuts at eye level only have a deep ridge in the hulls at this time. Nuts in the top southwest quadrant of the tree split first. Use a long pruning shear extension to cut small branches from this top portion of five or six trees in the orchard to check for initiation of hull split. Nuts on border trees will split first; check trees in the main part of the orchard to determine if hull split has begun. Sample for initiation of hull split twice a week. The hull does not need to be completely open to be considered split, just enough so that a visible opening is present without squeezing the nut (Stage 2, or disappearance of Stage 1; see Figure 6, page 12). Spraying at this time will provide a partially protective residue on the nuts. A large portion of the second generation eggs is laid within the split hull, protected from the hull split spray. However, a treatment applied at hull split provides about 40 to 60% control overall. This spray will reduce the number of early maturing, second generation moths and thus reduce the amount of egg laying that occurs in the early part of the third generation. Suppressing the early egg-laying period of the third generation will reduce the amount of navel orangeworm damage inflicted before nuts can be harvested.

A properly timed hull split spray will provide control of navel orangeworm and some control of peach twig borer. If you have a history of mite problems and mites are present in the orchard, use a miticide with the hull split spray.

STEPS	EXAMPLE
1. Expected delivered meat yield (and level of rejects[1]) for: Nonpareil lb/acre (%) Pollinizer 1 lb/acre (%) Pollinizer 2 lb/acre (%)	1500 (20)[2] 1200 (20) Merced 1500 (1) Mission
2. Penalties or premiums for reject level on Nonpareil, and pollinizers. (Based on a sliding penalty scale; obtain from your processor.)	−7¢/meat pound penalty for Nonpareil and Merced with 20% rejects. Applies to total meats delivered, including rejects
3. Expected good-meat price per pound.	$1.50
4. Gross revenue/acre (a) Each cultivar yield × reject level.[1]	(a) 1500 X .20 = 300 1200 X .20 = 240 1500 X .01 = 15
(b) Subtract reject pounds in (a) from cultivar yield to give good-meat yield.	(b) 1500 − 300 = 1200 1200 − 240 = 960 1500 − 15 = 1485
(c) Multiply good-meat yield by expected price for each cultivar.	(c) 1200 x 1.50 = 1800 960 x 1.50 = 1440 1485 X 1.50 = 2228
(d) Add premium or subtract penalty from Step 2 for each cultivar.	(d) 1800 − 105 = 1695 1440 − 84 = 1356 2228 − 0 = 2228
(e) Multiply results for each cultivar in (d) by fraction of the orchard represented by each cultivar.	(e) 1695 X .666 = 1129 1356 X .1625 = 220 2228 X .1625 = 362
(f) Sum results in (e) to give weighted average gross revenue/acre.	(f) 1129 + 220 + 362 = 1711
5. Added insect control practice for navel orangeworm.	Winter Cleanup of Mummies[3]

STEPS	EXAMPLE
6. Expected reduction in rejects.	50%
7. Level of rejects after control practice.	Nonpareil 10% Merced 10% Mission 1%
8. Gross revenue/acre after adding insect control practice. (a) Each cultivar yield × reject level from Step 7.	(a) 1500 X .10 = 150 1200 X .10 = 120 1500 X .01 = 15
(b) Subtract reject pounds in (a) from cultivar yield to give good-meat yield.	(b) 1500 − 150 = 1350 1200 − 120 = 1080 1500 − 15 = 1485
(c) Multiply good-meat yield by expected price for each cultivar.	(c) 1350 x 1.50 = 2025 1080 x 1.50 = 1620 1485 x 1.50 = 2228
(d) Add premium or subtract penalty for each cultivar. (Based on a sliding penalty scale; obtain from your processor.)	(d) (−2¢/lb penalty for Nonpareil and Merced with 10% rejects) 2025 − 30 = 1995 1620 − 24 = 1596 2228 − 0 = 2228
(e) Multiply results for each cultivar in (d) by fraction of the orchard represented by each cultivar.	(e) 1995 x .666 = 1329 1596 x .1625 = 259 2228 x .1625 = 362
(f) Sum results in (e) to give weighted average gross revenue/acre.	(f) 1329 + 259 + 362 = 1950
9. Subtract 4(f) from 8(f) to give revenue change due to new practice.	1950 − 1711 = 239
10. Added cost/acre of practice.	$34
11. Net benefits (losses)/acre from added practice. (Step 9 minus Step 10.)	$205

[1]Reject levels anticipated without insect control practices added in Step 6.

[2]Base expected cultivar yield and level of rejects used in Step 1 on previous crops, as well as any information relevant to the current year's crop.

[3]Obtain Steps 5 and 6 from Table 7.

Performing a second shake at harvest can increase your overall yield about 4 to 5%. To determine the cost effectiveness of a second shake, do Step 1-4 first with the increased yield, then without. Subtract the second from the first and proceed to Steps 10-11.

Figure 44. Worksheet for budgeting added costs and returns of almond insect management practices. (Adapted from "Economic analysis of navel orangeworm control in almonds" by Joseph C. Headley, in *California Agriculture*, May–June 1983, pp. 26–29.)

When application is timed to Nonpareil hull split, insecticides do not provide protection for the unsplit, soft-shelled cultivars in the same orchards. Leaving later maturing cultivars unsprayed at Nonpareil hull split will reduce insecticide costs and provide reservoirs for predators of pest mites. If you are treating for mites at the same time, however, the entire orchard must be sprayed. Monitor later maturing, soft-shelled cultivars closely for third generation navel orangeworm activity when the hulls on these cultivars begin to split. Crack the nuts and check for larvae. Consider a preharvest spray to protect these cultivars if damage seems likely. Remove all harvested nuts from the ground, however, before spraying later maturing cultivars to avoid violating preharvest intervals.

Evaluating Cost Effectiveness of Navel Orangeworm Control Practices. The usefulness of any crop management technique is determined ultimately by its cost effectiveness. Use the worksheet in Figure 44 and the information in Table 7 to determine your own costs and profits for each of the control recommendations.

Peach Twig Borer
Anarsia lineatella

Peach twig borer is a major pest of almonds, peaches, and stone fruits such as plums, prunes, and apricots. Originally from Europe, the pest was first reported in California in the 1880s and is now found throughout the state wherever its host plants are grown.

Until arrival of the navel orangeworm in the 1940s, peach twig borer was the most serious insect pest of almonds. Today there is a relationship between the two insects: nuts attacked by peach twig borer are preferred over sound nuts by navel orangeworm for egg laying.

Description

Moths of the peach twig borer are nocturnal and rest on the undersides of large limbs and in bark crevices during the day. The moths have steel gray, mottled forewings.

Each female lays about 90 to 100 eggs on twigs and fruit in spring and summer and on young branches in fall. Eggs are also laid on the undersides of leaves next to veins or the midrib.

When first laid, the oval eggs are white with a hexagonal pattern over the surface. As the embryo develops inside, the color changes to orange. Depending upon temperature, eggs hatch in 4 to 18 days.

When they first hatch, twig borer larvae are light brown with a black head and prothorax. As they grow, the head and prothorax remain black, but the body turns chocolate brown, and white portions between each body segment give the appearance of bands. The larvae have 4 to 5 instars (overwintering larvae have the additional instar) and take 10 days to 4 weeks to mature, depending on temperature.

Larvae are primarily twig borers, but they also feed on nuts. When they are ready to pupate, the larvae leave the shoot or nut they have been inhabiting, making an exit hole as they leave. The pupal stage lasts 4 to 11 days and usually occurs in crevices or curled leaves on the tree, debris on the ground, or for later generations, between the hull and shell. The pupae are dark brown and are not in a cocoon.

Seasonal Development

Peach twig borers overwinter as first or second instar larvae in cells (hibernacula) bored under the thin bark of limb crotches in 1- to 4-year-old wood. Larvae are inactive when it is cold but do not diapause. They may feed within hibernacula exposed to the sun on warm winter days. To locate a hibernaculum, look for minute, chimneylike piles of frass and sawdust that larvae cast out when they are constructing the cells. It is easier to see hibernacula in fall before the rains wash away the frass chimneys or in early spring when activity picks up and new particles of wood and frass are evident.

When the trees bloom in spring, larvae leave their hibernacula and feed on flower buds or young foliage. They can damage several clusters of leaves or buds before settling down and mining into the interior of a twig. The mine extends 1 to 2 inches from the shoot tip, and the affected portion of the twig wilts resulting in the characteristic damage symptom called "flagging."

There are usually four generations of peach twig borer each year in almonds. Overwintering larvae develop mostly in green shoots during spring, and most of the moths emerge from mid-April through May and lay first generation eggs. Larvae of the first generation develop in green shoots and occasionally in immature fruit. First generation moths emerge from mid-June through July and lay second generation eggs on both green nuts and leaf shoots. Figure 45 outlines the general seasonal development of peach twig borer. When second generation moths begin to emerge in August, they lay eggs on maturing nuts. Some of the larvae of this generation and the previous one can go into hibernacula for overwintering; others continue to develop on nuts or in shoots, and third generation moths emerge generally in October. These moths lay fourth generation eggs, and the larvae build hibernacula soon after hatching.

Adult peach twig borers are small, inconspicuous, gray moths.

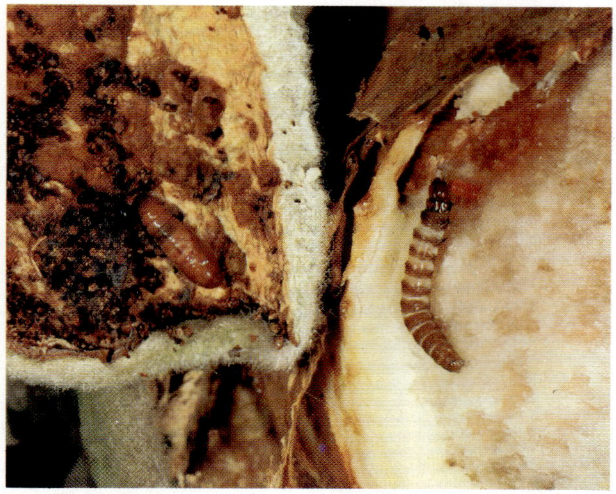

The twig borer larva is a small, distinctive caterpillar with a black head and thorax and a chocolate brown body with white portions between each body segment.

PEACH TWIG BORER

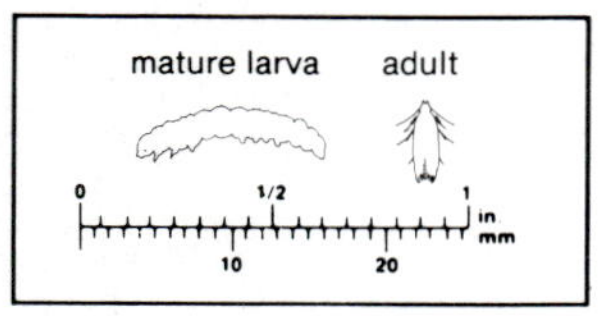

Peach twig borer eggs are white when first laid; as the embryo develops inside, the color changes to orange.

Hibernacula for peach twig borer are minute, chimneylike piles of frass and sawdust.

The overwintering site for twig borer larvae can be located by looking in the limb crotches of first- or second-year wood.

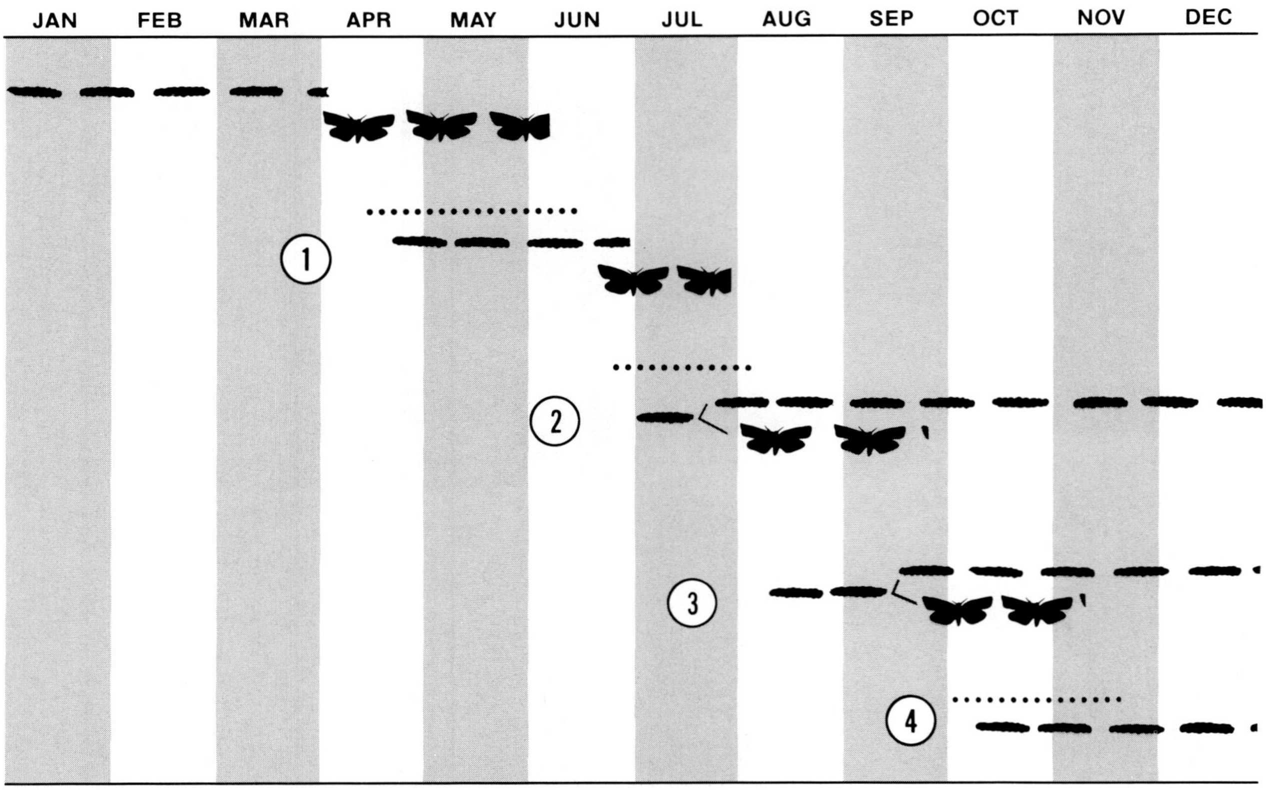

Figure 45. Seasonal development of the peach twig borer. The overwintering generation includes a portion of the second and third generation larvae, which went into hibernacula during the summer, and the fourth generation larvae. Timing of moth flights may vary from one year to the next. The moth symbols represent a general time period *during which* flights usually occur; actual flights generally do not last the entire length of time indicated by the symbols.

Damage

Peach twig borer larvae can damage both the growing shoots and the nut crop. They mine the tender shoots of the almond tree, killing the tips and causing lateral twig growth. This can be serious in young trees because shoot damage can make it difficult to train the trees during pruning. On older trees this type of damage is unimportant.

Oriental fruit moth larvae also mine the growing tips of shoots but have white or pink bodies with brown head capsules in contrast to the distinctly segmented, dark brown bodies and black head capsules of peach twig borers. They also mine deeper into the shoots than the peach twig borer.

Like the navel orangeworm, twig borer larvae can feed on the hulls, between the hull and shell, or directly on the nutmeats; but twig borer larvae differ greatly in appearance from navel orangeworm larvae and do not produce webbing. Also, navel orangeworms bore into the nutmeat and several larvae may feed in a single nut, whereas peach twig borers cause only shallow channels and surface grooves on the kernels and normally feed one larva to a nut. Peach twig borer and navel orangeworm often infest the same nut. Because peach twig borer feeding is often masked by navel orangeworm feeding, peach twig borer infestations are usually greater than that detected in the harvest sample.

Nuts of certain cultivars seem to be more severely damaged by twig borer than others. Susceptibility to peach twig borer damage is generally highest in the Merced and Thompson cultivars, followed by Nonpareil, Fritz, Price, and Ne Plus Ultra.

Management Guidelines

Biological Control. Peach twig borer has about 30 species of natural enemies. Among those most commonly found in California are the chalcid wasps, *Paralitomastix varicornis* and *Hyperteles lividus*, and the grain or itch mite,

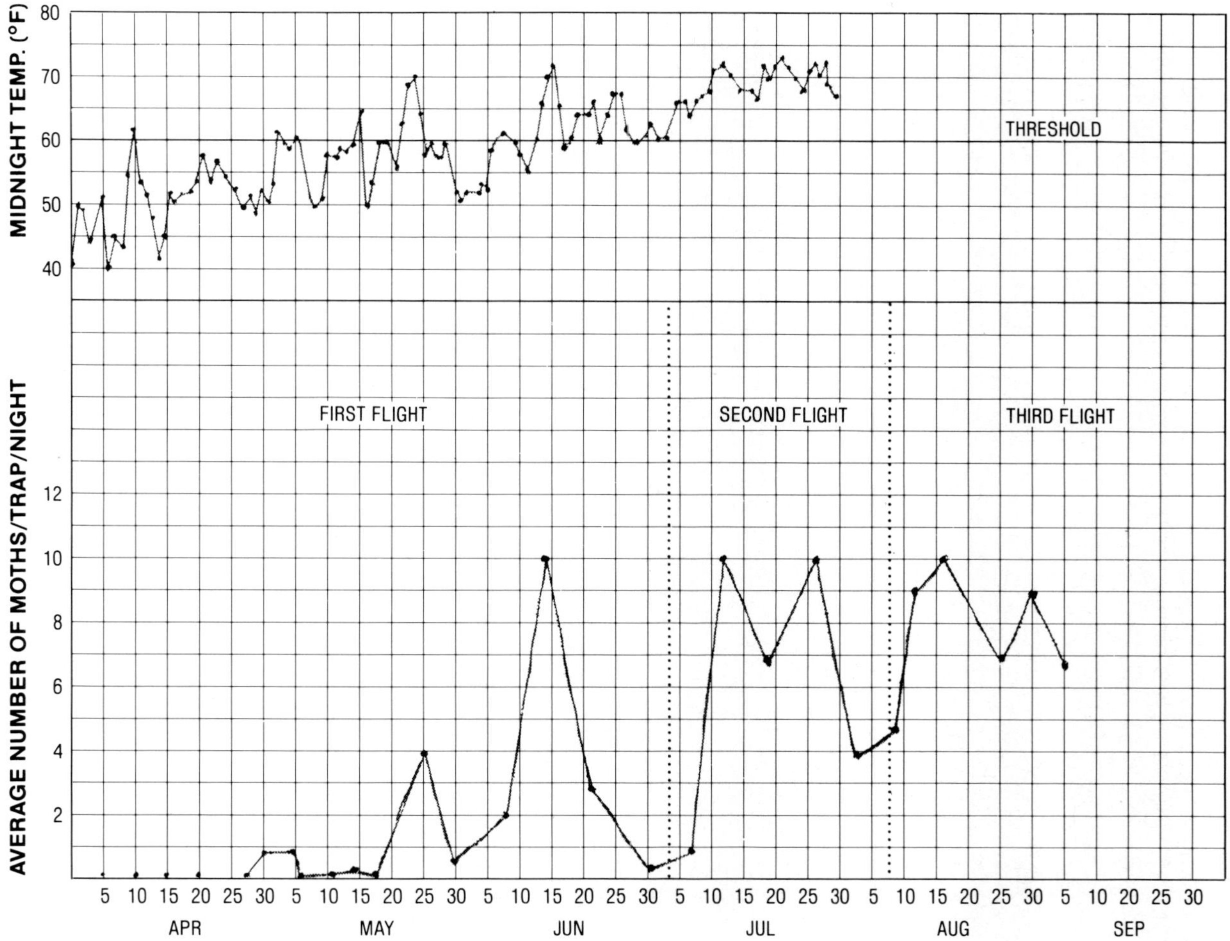

Figure 46. Graph of peach twig borer pheromone trap catches and daily midnight temperatures. Note that trap catches decline when temperatures fall below the threshold of 60°F.

Pyemotes ventricosus. In some years these natural enemies destroy a significant portion of larvae, but by themselves they do not seem to play an important role in reducing twig borer populations below economically damaging levels. No effective cultural controls for peach twig borer are known.

Monitoring and Control. Sprays for peach twig borer can be applied during dormancy or during the growing season: in spring or at hull split. The most effective time to spray is during dormancy.

Dormant Sprays. Treat your orchards every year for peach twig borer with a dormant spray of oil and organophosphate insecticide. Spraying during tree dormancy has several advantages: coverage of the bark is more complete; beneficial insects and mites are less affected; other pests such as San Jose scale, brown mite, and European red mite are simultaneously controlled; and there are fewer conflicts with other cultural operations in the orchard. The dormant spray kills overwintering twig borer larvae in their hibernacula. If you properly apply the spray, twig borer populations can be reduced over 95%. Avoid spraying when almond trees or orchard cover crops are in bloom and bees are present in the orchard. Also, be sure to read the pesticide label and observe the minimum interval between spraying and bringing bees into the orchard.

Spring Sprays. Specific guidelines on how to determine if a spring spray is required are not available. If you did not spray during the dormant season and you have a history of peach twig borer damage, you will need to apply a spring spray. In most years this should effectively control peach twig borer until the crop is harvested.

This shoot terminal has been mined by a twig borer larva, and the affected portion of the twig has wilted, causing a symptom known as "flagging."

Twig borer feeding on almonds causes shallow channels and surface grooves on the kernels.

The spring spray is directed against the hatching larvae of the first generation. If your control emphasis is peach twig borer, time the spray using sex pheromone traps and degree-day calculations. If control emphasis is navel orangeworm or both navel orangeworm and peach twig borer, time the spray based on data collected from monitoring navel orangeworm egg traps.

Place pheromone traps in the orchard by March 20 in the southern portions of the Central Valley and April 1 in the northern areas to detect moth emergence. Use an average of one trap per 20 acres to monitor the orchard; place the traps at least one acre (209 ft) apart (see Monitoring Methods, page 50 for additional information). Hang the traps 6 to 7 feet (2 m) high in the northeast quadrant of the tree, 1 to 3 feet (0.3 to 1 m) from the outside of the canopy. Count and remove the moths at least twice weekly during major flight activity. Record and graph the average number of moths per trap per night (Figures 40, page 52, and 46). Replace pheromone dispensers every 2 weeks and trap bottoms when they become obviously dirty or after 200 to 250 trapped moths have been removed.

Flight activity of the peach twig borer is greatest between the hours of midnight and 2 a.m. and at temperatures above 60°F (15.5°C). When temperatures fall below this threshold, or if it's windy, very little flight activity takes place. Pay close attention to nightly temperatures if your trap data indicates a major flight is occurring. If the temperature drops below 60°F at midnight, your trap

Date	Daily Low	Daily High	Degree-Days for Date	Total Degree-Days Following Oviposition
4/10	47	72	10.0	10.5
4/11	46	73	10.5	20.5
4/12	46	75	11.5	32.0
4/13	47	74	11.0	43.0
4/14	48	74	11.0	54.0
4/15	46	75	11.5	65.5
4/16	46	76	12.0	77.5

Figure 47. Example of degree-day accumulations for peach twig borer to determine the optimum time for a May spray. Add together the total number of degree-days that accumulate after the April flight begins; spray after 400 to 500°D have accumulated.

catches will decline, but they will increase when the weather turns warm (Figure 46). This increase does not necessarily signify the start of the next peach twig borer generation, it could be the resumption of the interrupted one. Rely on degree-day calculations to determine when the next generation has started.

Begin calculating degree-days in the spring when male moths are detected in the traps on at least two consecutive sampling dates. Use Table 8 to calculate peach twig borer degree-days from weather data. Add together the degree-days for each day beginning with the first moth catch as shown in Figure 47. About 220°D accumulate from first male moth catch to first egg hatch in the next generation. Optimum timing for first generation or spring larval treatment, called the "May spray," is between 400° and 500°D after the start of the spring flight in April.

Hull Split Sprays. Control of peach twig borer at hull split is difficult and not generally recommended because the timing is so critical. The hull split spray is directed at the hatching larvae of the second generation. To be effective, it must be applied at very early hull split: when about 1% of the hulls of sound nuts in the treetops have split.

LOWER THRESHOLD: 50.00 F UPPER THRESHOLD: 88.00 F SINGLE SINE METHOD HORIZONTAL CUTOFF

MINIMUM TEMPERATURES

MAX TEMPS	88	86	84	82	80	78	76	74	72	70	68	66	64	62	60	58	56	54	52	50	48	46	44	42	40	38	36	34
118	38	38	37	37	36	36	35	35	34	33	32	32	31	30	29	28	28	27	26	25	24	24	23	23	22	22	21	21
116	38	38	37	37	36	36	35	34	34	33	32	31	31	30	29	28	27	27	26	25	24	24	23	23	22	22	21	21
114	38	38	37	37	36	36	35	34	34	33	32	31	31	30	29	28	27	26	26	25	24	23	23	22	22	21	21	21
112	38	38	37	37	36	36	35	34	34	33	32	31	30	30	29	28	27	26	25	24	24	23	22	22	22	21	21	20
110	38	38	37	37	36	36	35	34	33	33	32	31	30	29	28	28	27	26	25	24	23	23	22	22	21	21	20	20
108	38	38	37	37	36	35	35	34	33	32	32	31	30	29	28	27	27	26	25	24	23	22	22	21	21	20	20	20
106	38	38	37	37	36	35	35	34	33	32	31	31	30	29	28	27	26	25	24	24	23	22	22	21	21	20	20	19
104	38	38	37	37	36	35	34	34	33	32	31	30	30	29	28	27	26	25	24	23	22	22	21	21	20	20	19	19
102	38	38	37	37	36	35	34	34	33	32	31	30	29	28	27	27	26	25	24	23	22	21	21	20	20	19	19	19
100	38	38	37	36	36	35	34	33	32	32	31	30	29	28	27	26	25	24	23	22	22	21	20	20	19	19	19	18
98	38	38	37	36	36	35	34	33	32	31	30	30	29	28	27	26	25	24	23	22	21	21	20	19	19	19	18	18
96	38	38	37	36	35	35	34	33	32	31	30	29	28	27	26	25	24	23	23	22	21	20	20	19	19	18	18	17
94	38	38	37	36	35	34	33	33	32	31	30	29	28	27	26	25	24	23	22	21	20	20	19	18	18	18	17	17
92	38	37	37	36	35	34	33	32	31	30	29	28	27	26	25	24	23	22	21	20	20	19	18	18	17	17	17	16
90	38	37	36	36	35	34	33	32	31	30	29	28	27	26	25	24	23	22	21	20	19	18	18	17	17	16	16	16
88	38	37	36	35	34	33	32	31	30	29	28	27	26	25	24	23	22	21	20	19	18	18	17	16	16	16	15	15
86		36	35	34	33	32	31	30	29	28	27	26	25	24	23	22	21	20	19	18	17	17	16	15	15	15	14	14
84			34	33	32	31	30	29	28	27	26	25	24	23	22	21	20	19	18	17	16	16	15	15	14	14	13	13
82				32	31	30	29	28	27	26	25	24	23	22	21	20	19	18	17	16	15	15	14	14	13	13	12	12
80					30	29	28	27	26	25	24	23	22	21	20	19	18	17	16	15	14	14	13	13	12	12	11	11
78						28	27	26	25	24	23	22	21	20	19	18	17	16	15	14	13	13	12	12	11	11	11	10
76							26	25	24	23	22	21	20	19	18	17	16	15	14	13	12	12	11	11	10	10	10	9
74								24	23	22	21	20	19	18	17	16	15	14	13	12	11	11	10	10	9	9	9	8
72									22	21	20	19	18	17	16	15	14	13	12	11	10	10	9	9	8	8	8	8
70										20	19	18	17	16	15	14	13	12	11	10	9	9	8	8	8	7	7	7
68											18	17	16	15	14	13	12	11	10	9	8	8	7	7	7	6	6	6
66												16	15	14	13	12	11	10	9	8	7	7	6	6	6	6	5	5
64													14	13	12	11	10	9	8	7	6	6	5	5	5	5	4	4
62														12	11	10	9	8	7	6	5	5	5	4	4	4	4	4
60															10	9	8	7	6	5	4	4	4	3	3	3	3	3
58																8	7	6	5	4	3	3	3	3	2	2	2	2
56																	6	5	4	3	2	2	2	2	2	2	1	1
54																		4	3	2	2	1	1	1	1	1	1	1
52																			2	1	1	1	0	0	0	0	0	0
50																				0	0	0	0	0	0	0	0	0
48																					0	0	0	0	0	0	0	0

Table 8. Degree-Day Table for Peach Twig Borer Based on a Lower Threshold of 50°F and an Upper Threshold of 88°F. For each day, locate the low and high temperatures on the table and follow the column and row to where they intersect to find the total degree-days. For odd-numbered temperatures, interpolate between numbers.

San Jose Scale
Quadraspidiotus perniciosus

San Jose scale was accidentally introduced into California during the 1870s from China. It now is widespread and known to infest over 700 species of plants. The most susceptible of these hosts are deciduous fruit and nut trees such as pears, plums, apples, peaches, nectarines, and almonds. Severe infestations will kill fruit wood, reduce yields, and eventually kill the tree.

Description and Seasonal Development

San Jose scales emerge as nymphs; hence there is no visible egg stage. Three stages occur during the first instar: the "crawler," the "white cap," and the "black cap." The crawler stage is the only one in which a scale infestation can spread. All other developmental stages, except adult males, are immobile and attached to a single feeding site under a protective scale covering.

The bright yellow crawler is about the size of the sharp end of a pin (0.2 mm) and resembles a mite. After leaving the protection of the female's cover, it relocates by crawling, by being carried on equipment or on birds or insects that land on branches where the crawlers are active, or by wind. Within 8 to 24 hours of birth it settles down on the bark, inserts its mouth parts into the host, and begins to feed on the tree's sap. As it feeds, a white, waxy covering (the white cap) is secreted. After about a week, a band of dark wax appears around the periphery of the white cap, marking the beginning of the black cap stage. Eighty percent of San Jose scale overwinter in the black cap stage; during mild winters, the remaining 20% are mostly mated adult females. These females will not survive a severe winter, however.

When the weather warms in February and the tree's sap flows, black cap nymphs resume their growth and undergo the first molt. The male's scale covering elongates in the latter part of the second instar, and they become distinguishable from females, whose covering remains circular. Females molt twice before reaching maturity, while males molt a total of four times.

Following the second molt, males pass through two nonfeeding instars in which the scale cover remains elongated, is gray, and has a small, indistinct knob near the larger end. After the fourth molt, the male emerges from under the scale cover as a delicate, minute, two-winged, yellow brown insect. Adult males can mate immediately after emerging and most only live 6 to 8 hours in warm weather.

Females live under the scale covering for the remainder of their lives. At maturity the cover is gray brown and has a tiny white knob in the center. The insect body under the covering is bright yellow and has no appendages.

When females are mature, they produce a sex pheromone to attract males. The males' spring flight period coincides with the receptivity of the females in March and April. Within 5 to 6 weeks, females begin producing first generation crawlers. Depending upon temperature, a female can produce crawlers for 6 to 8 weeks at a rate of about 10 per day.

Crawlers that emerge in May give rise to the first generation male flight in June. Two more generations follow in July-August and September-October. Crawlers produced in October and November overwinter as black caps and produce the overwintering flight the following spring. Under optimum conditions of hot, dry weather, a generation can be completed in 7 to 8 weeks.

Damage

San Jose scales suck plant juices from the trees and inject a toxin, contributing to death of twigs and limbs and to overall decline in tree vigor, growth, and productivity. The scale feeds on any thin-barked wood. On 1-year-old wood, it produces a red halo around the feeding site. San Jose scale is usually not found on almond hulls. Infested trees appear water stressed, and extremely heavy populations can cause the bark to crack and exude gum. Branches may be killed, permanently affecting structure, vigor, and productivity of mature orchards. In younger trees, untreated infestations can kill the tree within 1 to 3 years.

Management Guidelines

Biological Control. Natural enemies that feed on San Jose scale include two predaceous beetles: the twicestabbed lady beetle, *Chilocorus ortus* and another small beetle *Cybocephalus californicus.* A number of small chalcid and aphelinid wasps parasitize this scale. These predators and parasites may be helpful in reducing scale populations; but insecticides used during the growing season for other pests disrupt this natural control, and scale populations often increase rapidly. Low winter mortality due to mild temperatures will also permit a buildup of scale populations.

Monitoring and Control. Examine the orchard during dormancy to spot San Jose scale infestations. Twigs or branches that have been killed by the scale retain their leaves during the winter. Scale populations are often irregularly distributed within orchards and may be hidden beneath loose bark in older trees. Frequently, infestations first appear in the tops of trees where it is hardest to get good spray coverage. Examine prunings, especially those from the treetops, to detect infestations. Use pheromone traps or sticky tapes (see section on spring sprays below)

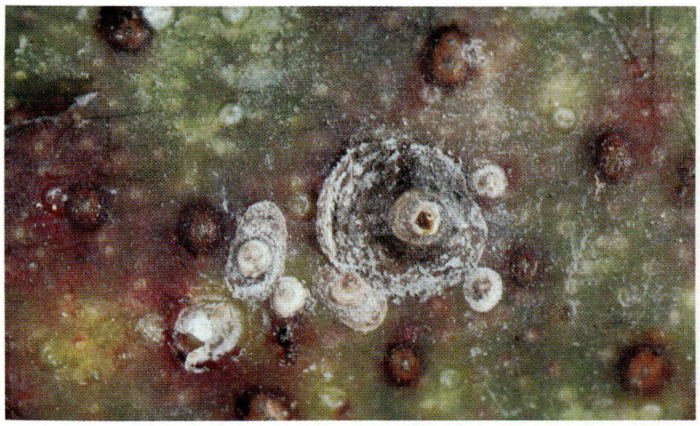

Following the first molt, the male San Jose scale's covering is elongated, while the female's remains circular. White cap and black cap stages can also be seen in this photo.

If you remove the covers from San Jose scales, you can see the bright yellow bodies of the male and female underneath.

Tiny, bright yellow crawlers of the San Jose scale migrate to new feeding locations. After crawlers settle down to feed, they secrete a white, waxy covering, known as the white cap stage.

The adult male San Jose scale has a dark band across its back (left). *Aphytis melinus* are tiny, yellow wasps that parasitize San Jose scale. They resemble the adult male scale, but lack the dark band across the back (right).

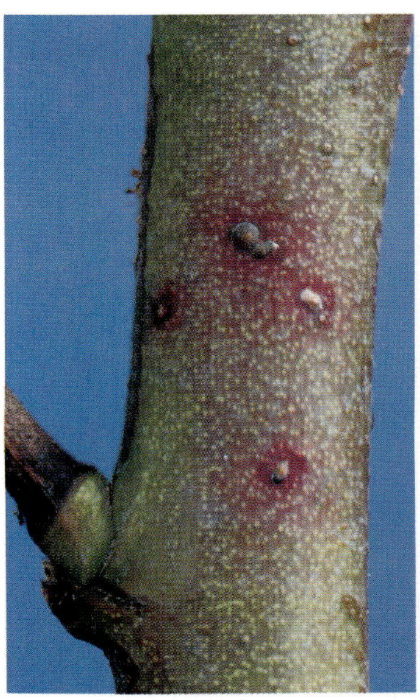

Red halo around the feeding site is evidence of San Jose scale on young wood.

Use a pheromone trap to monitor flight activities of the adult male San Jose scale.

SAN JOSE SCALE

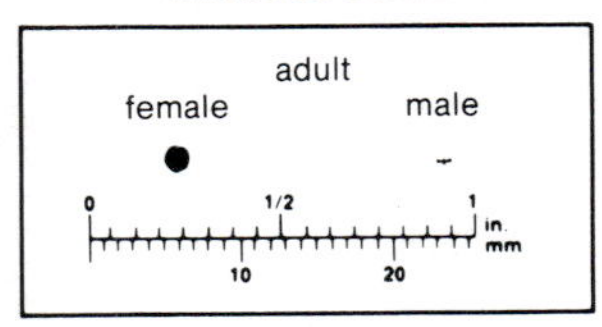

TWICE-STABBED LADY BEETLE

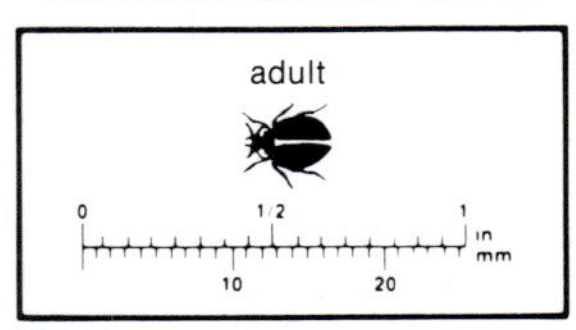

CYBOCEPHALUS

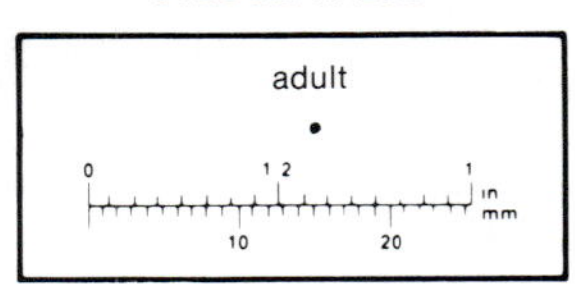

to detect San Jose scale infestation during the growing season and to time a spring spray, if needed.

Dormant Sprays. Each year, spray every orchard during dormancy to control or prevent flare-ups of San Jose scale. Scale occurs in dense colonies, so good spray coverage is important. If dormant sprays are omitted or improperly applied, damage can be expected within 1 to 3 years. Properly applied dormant sprays eliminate the spring flight and suppress the infestation thoughout the growing season.

Spring Sprays. If a dormant spray is not applied, the best alternate time to spray is during crawler emergence in May in the southern portion of the state, and in late May/early June in northern areas. Do not try to control San Jose scale and navel orangeworm with the same spray, however. Not only are they best controlled by different materials, but the spray timing may differ.

Seasonal flights of male scales can be monitored with pheromone traps. Place the tent-shaped traps 6 to 7 feet (2 m) high in the north or east sides of the trees by February 25 in southern districts and by March 15 in northern districts. Securely fasten them to tree limbs with a wire hanger. This will prevent traps from being blown out of the trees by wind. Use at least three to four traps, regardless of orchard size. In orchards larger than 60 acres, use one trap per 20 acres (see the section on Monitoring Methods, page 50 for additional information).

Inspect the traps twice weekly for male scales. Because male scales are so small, use a hand lens to inspect the traps. Look for the dark band on the back of the male scale to distinguish it from another small yellow insect, *Aphytis* spp., which is a parasite of the scale. Record your data on a form similar to the one in Figure 40, and clean the trap. Replace traps whenever they become too dirty for accurate identification and counting of male scales; replace pheromone caps every 4 weeks.

LOWER THRESHOLD: 51.00 F UPPER THRESHOLD: 90.00 F SINGLE SINE METHOD HORIZONTAL CUTOFF

MINIMUM TEMPERATURES

MAX TEMPS	90	88	86	84	82	80	78	76	74	72	70	68	66	64	62	60	58	56	54	52	50	48	46	44	42	40	38	36	34
118	39	39	38	38	37	37	36	35	35	34	33	32	32	31	30	29	28	28	27	26	25	24	24	23	23	22	22	22	21
116	39	39	38	38	37	37	36	35	35	34	33	32	32	31	30	29	28	27	27	26	25	24	24	23	23	22	22	21	21
114	39	39	38	38	37	37	36	35	35	34	33	32	31	31	30	29	28	27	26	25	25	24	23	23	22	22	21	21	21
112	39	39	38	38	37	37	36	35	34	34	33	32	31	30	29	29	28	27	26	25	24	24	23	22	22	22	21	21	20
110	39	39	38	38	37	36	36	35	34	33	33	32	31	30	29	28	28	27	26	25	24	23	23	22	22	21	21	20	20
108	39	39	38	38	37	36	36	35	34	33	32	32	31	30	29	28	27	26	25	25	24	23	22	22	21	21	20	20	20
106	39	39	38	38	37	36	35	35	34	33	32	31	31	30	29	28	27	26	25	24	23	23	22	21	21	21	20	20	19
104	39	39	38	38	37	36	35	35	34	33	32	31	30	29	28	28	27	26	25	24	23	22	22	21	21	20	20	19	19
102	39	39	38	37	37	36	35	34	33	33	32	31	30	29	28	27	26	25	24	23	23	22	21	21	20	20	19	19	19
100	39	39	38	37	37	36	35	34	33	32	31	31	30	29	28	27	26	25	24	23	22	21	21	20	20	19	19	18	18
98	39	39	38	37	36	36	35	34	33	32	31	30	29	28	27	26	25	24	24	23	22	21	20	20	19	19	18	18	18
96	39	39	38	37	36	35	34	34	33	32	31	30	29	28	27	26	25	24	23	22	21	20	20	19	19	18	18	17	17
94	39	38	38	37	36	35	34	33	32	31	30	29	28	27	26	25	24	23	22	21	21	20	19	19	18	18	17	17	17
92	39	38	37	37	36	35	34	33	32	31	30	29	28	27	26	25	24	23	22	21	20	19	19	18	17	17	17	16	16
90	39	38	37	36	35	34	33	32	31	30	29	28	27	26	25	24	23	22	21	20	19	18	18	17	17	16	16	15	15
88		37	36	35	34	33	32	31	30	29	28	27	26	25	24	23	22	21	20	19	18	17	17	16	16	15	15	15	14
86			35	34	33	32	31	30	29	28	27	26	25	24	23	22	21	20	19	18	17	16	16	15	15	14	14	14	13
84				33	32	31	30	29	28	27	26	25	24	23	22	21	20	19	18	17	16	15	15	14	14	13	13	13	12
82					31	30	29	28	27	26	25	24	23	22	21	20	19	18	17	16	15	14	14	13	13	12	12	12	11
80						29	28	27	26	25	24	23	22	21	20	19	18	17	16	15	14	13	13	12	12	12	11	11	11
78							27	26	25	24	23	22	21	20	19	18	17	16	15	14	13	12	12	11	11	11	10	10	10
76								25	24	23	22	21	20	19	18	17	16	15	14	13	12	11	11	10	10	10	9	9	9
74									23	22	21	20	19	18	17	16	15	14	13	12	11	10	10	9	9	9	8	8	8
72										21	20	19	18	17	16	15	14	13	12	11	10	9	9	9	8	8	8	7	7
70											19	18	17	16	15	14	13	12	11	10	9	8	8	8	7	7	7	6	6
68												17	16	15	14	13	12	11	10	9	8	8	7	7	6	6	6	6	5
66													15	14	13	12	11	10	9	8	7	7	6	6	5	5	5	5	5
64														13	12	11	10	9	8	7	6	6	5	5	5	4	4	4	4
62															11	10	9	8	7	6	5	5	4	4	4	4	3	3	3
60																9	8	7	6	5	4	4	3	3	3	3	3	2	2
58																	7	6	5	4	3	3	2	2	2	2	2	2	2
56																		5	4	3	2	2	2	1	1	1	1	1	1
54																			3	2	1	1	1	1	1	1	1	1	1
52																				1	0	0	0	0	0	0	0	0	0
50																					0	0	0	0	0	0	0	0	0
48																						0	0	0	0	0	0	0	0

Table 9. Degree-Day Table for San Jose Scale Based on a Lower Threshold of 51°F and an Upper Threshold of 90°F. To find total degree-days for a day, locate the low and high temperatures and follow the column and row to where they intersect. For odd-numbered temperatures, interpolate between numbers.

After the first male scales have been trapped, begin calculating degree-days to predict crawler emergence, using the degree-day table (Table 9) and adding degree-days for each day after the first male scale catch in the same manner as shown in Figure 47 for the peach twig borer. Crawler emergence begins about 405°D after the first male flight. Optimum timing for a May crawler spray is 600° to 700°D after the first male scale is collected in March/April.

Researchers are currently investigating the use of sticky tape traps to detect San Jose scale. Initial experiments indicate these traps may be more effective in detecting the spring generation of San Jose scale than pheromone traps. Because male scales are weak flyers, erratic spring weather can greatly influence their flight activity and, thus, the effectiveness of pheromone traps. Sticky tape traps consist of clear plastic tape that is wrapped around scaffold limbs. Emerging males and crawlers that are migrating to a new location stick to the surface of the tape. Optimum spray timing using sticky tape traps is 200° to 300°D after the first crawlers are trapped. Consult your farm advisor for additional information.

If a late season scale population has flared up in your orchards because the previous dormant spray was missed or improperly applied, do not try to treat the infestation during the growing season, except during the May crawler period. If this period has passed, wait until the following dormant season to apply a spray. Late fall or postharvest treatments are not effective against San Jose scale because of overlapping generations. Only crawlers, white caps, black caps, and emerging males are susceptible to sprays. Also, treatments at these times are very disruptive to natural enemies of mites and other insect pests.

Olive Scale
Parlatoria oleae

The olive scale's covering is similar in shape, size, and color to San Jose scale, but the insect's body underneath is purple. Olive scale is not widely distributed in almonds but occurs in localized areas in some counties. Like San Jose scale, it secretes a tough, waxy covering that both protects and camouflages the insect. In spring, olive scale lays eggs beneath its shell. Eggs begin to hatch in late April and crawlers scatter to new sites. If undisturbed by insecticides applied for major almond pests, natural enemies provide good control of olive scale. This scale is also adequately controlled by the oil and organophosphate applied in the dormant spray.

European Fruit Lecanium
Lecanium corni

European fruit lecanium, formerly called brown apricot scale, occurs throughout the Central Valley but is rarely a problem. It passes the winter as an unprotected nymph. Control is best at this stage (before February 1) because in late winter females harden, exude a rounded shell, and are no longer effectively controlled with pesticides. The domed shell is shiny brown, about 3/8-inch (10 mm) in diameter, and has several ridges along the back. Eggs are laid in the spring, and the adult female's vital organs waste away as eggs accumulate under the arched canopy of the shell. Hatching occurs from May to July; there is only one generation a year. The young develop through the remainder of the season and produce copious

The olive scale is purple underneath its covering. The color helps distinguish it from San Jose scale, which is yellow.

The domed, shiny brown shell of the European fruit lecanium has several ridges along the back.

amounts of honeydew. Although the nymphs can be controlled by insecticides once they have emerged from under the female's domed cover, wait until winter when the leaves are off the trees for best coverage and control with a dormant spray.

MITES

Six species of mites have the potential to cause economic damage in Central Valley almond orchards. Their pest status depends upon various cultural practices, weather, and the level of predation against them.

Two of these mite species, twospotted and Pacific spider mites, occur throughout the Central Valley. Strawberry spider mites are occasionally found on almond trees in areas south of Merced County. In the early season, all three of these spider mites, collectively referred to in this manual as webspinning mites, are found in the lower central portion of the tree; once the weather gets hot, their numbers grow rapidly, and they disperse throughout the tree.

European red mites and brown mites are generally found in the northern San Joaquin and Sacramento valleys; however, European red mites can also occur in high numbers in the central and southern San Joaquin Valley. Both mites are found on leaves throughout the tree canopy and are potentially damaging early in the growing season. Populations of European red mites may remain high throughout the growing season.

Citrus red mites occur in the southern San Joaquin Valley and frequently are found in almond orchards near citrus groves. In spring, citrus red mites are carried in by wind from citrus groves and are first found mostly in the upper periphery of the almond trees. Citrus red mites also have been found, on occasion, in almond orchards far removed from citrus plantings. Populations of citrus red mites grow rapidly in late spring and early summer.

Most of these mites overwinter in almonds. European red mites and brown mites overwinter as red eggs that are laid on spurs and in limb crotches. Twospotted, Pacific, and strawberry mites overwinter as red or orange adult females on ground litter, weeds, or under bark scales of the tree. Citrus red mites overwinter in citrus groves and reinvade almond orchards each spring.

Figure 48 illustrates the periods of activity for these pest mites.

Webspinning Mites

Pacific Spider Mite ***Tetranychus pacificus***
Twospotted Spider Mite ***Tetranychus urticae***
Strawberry Spider Mite ***Tetranychus turkestani***

Pacific, twospotted, and strawberry mites are webspinning mites. Dense webbing occurs on leaves and branch terminals when populations of these mites reach high numbers. Adults of these species are difficult to distinguish from each other in the field. All three may occur in the same orchard, but it is not known if they differ in the effect their feeding has on almond trees or if different almond cultivars vary in their susceptibility to damage by these mites. Because they all have similar life histories and are controlled in the same manner, however, there is no practical reason at this time to distinguish the three species when monitoring orchards.

Description and Seasonal Development

Webspinning mites overwinter as red or orange colored females under rough bark scales of almond trees and in ground litter and trash. During the growing season the color of the oval-shaped females ranges from light yellow green to almost black, depending upon age and host food. All have dark spots on both sides of the body. Adult male mites are smaller and less common than female mites. Male mites do not overwinter.

When the weather warms in the spring, females become active, begin feeding on young almond leaves, and lay their eggs on the undersides of leaves. The spherical eggs are translucent when first laid but turn opaque before hatching.

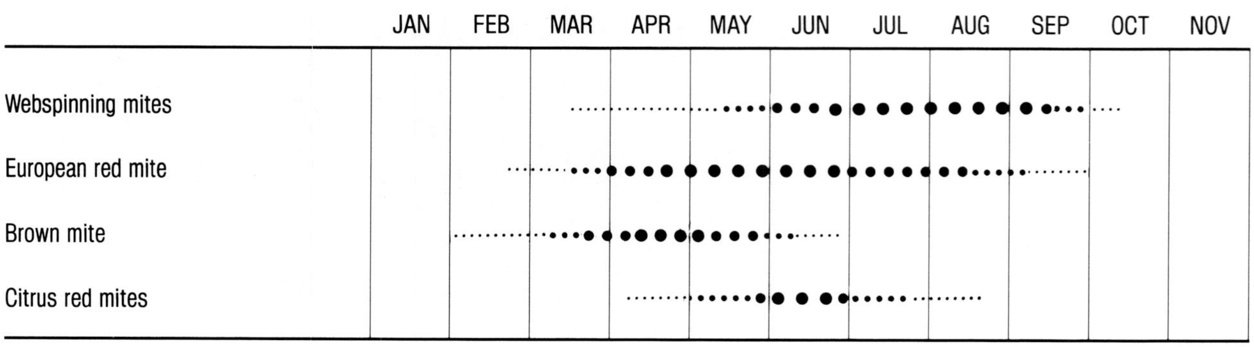

Figure 48. Typical periods of activity for pest mites in almond orchards. The larger the circle, the greater the potential for mite activity.

Immature mites molt three times. First stage mites have only six legs, but later stages and adults have eight. Early in the growing season, webspinning mites are typically found on foliage in central or lower portions of trees as they radiate out from overwintering sites. Spring populations are often localized in "hot spots" either along roadsides or on the orchard's periphery. They may also occur in high numbers in early spring in orchards that received in-season pyrethroid applications the previous season. During summer when the weather is hot and dry, mite infestations are common in dusty areas of an orchard, when mite predators are killed off by certain insecticides, or when predators cannot control them adequately in warmer than average years.

All three mites reproduce rapidly in hot weather and commonly become numerous in June through September. If temperature and food supply are favorable, a generation can be completed in 7 days or less, and 8 to 15 generations can occur during a single growing season. As foliage quality declines in heavily infested trees, mites are dispersed by wind currents to new locations. High mite populations generally undergo a rapid decline in late August or September when predation overtakes them, when host plant conditions deteriorate, and as the daylight hours decrease and the weather turns somewhat cooler. With the change in the quality of the host plant, daylight hours, and weather, females enter a period of diapause for the winter.

Damage

Mites cause damage by sucking cell contents from leaves. Initially, they cause a stippling effect on the leaves; as feeding continues, stippling becomes more severe and the leaves turn light yellow and fall. When spider mites are abundant, they often cover tree terminals with copious amounts of webbing.

Webspinning mites have the greatest potential for damage from late June through August; high populations at this time can result in partial to complete defoliation. Leaves that remain on the trees will not recover from mite damage after mite control has been achieved.

No effect can be detected on tree growth and productivity during the season of infestation and defoliation. This is because annual growth and nut sizing are complete by the time mite infestations become serious. The following season, however, nut set, leaf size, extension of terminals, and total crop may be significantly lowered; crop reduction can approach 20%.

Trees under water stress will show mite damage more readily than well-irrigated trees. Transpiration and photosynthesis rates are depressed more under the combined stress of water shortage and mite leaf feeding than when either factor is operating alone.

Overwintered female webspinning mites have red or orange bodies with dark spots on both sides of the body.

SPIDER MITES

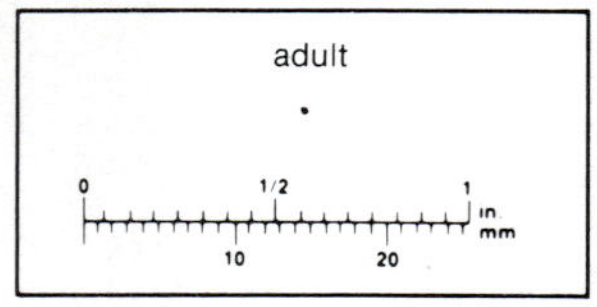

As the season progresses, webspinning mite adults are generally light colored with dark spots on both sides of the body, but their color depends on the quality of foliage. Eggs are spherical and translucent. The smaller mite in the center left is a male.

Feeding by webspinning mites first appears as a light stippling on the leaves, as seen in the leaf on the left side of this photo. The leaf in the center has moderate stippling.

When they are abundant, webspinning mites can produce copious amounts of webbing.

RICHARD E. RICE

Defoliation by webspinning mites not only affects the following year's yield but also creates a trash problem at harvest.

Even though the potential for mite damage is great, this potential is reached only when pest mite populations get out of hand. Low pest mite populations are necessary to maintain predator population in the orchard. Research indicates that until webspinning mites exceed a threshold of an average of 100 to 120 mite-days per leaf over the entire tree (one mite-day = one mite feeding on a leaf for one day) on *well-irrigated* trees, there is little effect on the tree. Therefore, effects from an average of four to five mites per leaf for 20 to 25 days may not be detectable. Because well-irrigated trees can tolerate feeding by more mites than water-stressed trees, the threshold for mite feeding in water-stressed trees will be lower.

Management Guidelines

Prevention. Good orchard management practices can reduce mite problems. For instance, dusty conditions often lead to mite outbreaks. Minimize dust by oiling orchard roads or watering roadways with a tank truck at regular intervals throughout the growing season. Maintaining a ground cover in the orchard will also help reduce dust. In problem areas of the orchard, limit the amount of traffic and always drive slowly on orchard roads.

A key aspect of mite management is adequate irrigation. Because mites cause greater damage to water-stressed trees, keep all trees in the orchard well irrigated and monitor mite populations carefully. Unless mites are under satisfactory control, withdrawing irrigation water and drying orchards in preparation for harvest will intensify defoliation related to mite damage.

Biological Control. Several predators play an important role in regulating mite populations. Orchards that do not receive in-season sprays may have several species of predatory mites including the western predatory mite (*Metaseiulus (=Typhlodromus) occidentalis*), as well as *Amblyseius (Euseius) hibisci*, *Typhlodromus caudiglans*, and *Typhlodromus citri*. In orchards that receive moderate applications of organophosphate insecticides, the predominant predatory mite is the western predatory mite.

The western predatory mite is one of the most effective mite predators. Although it prefers the egg stage, it feeds on all stages of the pest mites. It is about the same size as the webspinning mite but lacks dark spots on the sides of its body. The color of its body may range from cream to amber to red, depending on what it has been feeding on, but it never attains the deep red color of the citrus red or European red mite. Western predatory mite eggs are larger than those of webspinning mites and oval in shape, whereas webspinning mite eggs are spherical.

Western predatory mites can complete a generation in 7 days under optimal conditions and build up rapidly enough to regulate webspinning mites. To be effective, however, western predatory mites must be present in the orchard by May and early June so that sufficient numbers are available to respond to the rapid pest mite population increases associated with the onset of hot weather. If pest mites greatly outnumber the predatory mites, a better balance can often be achieved by applying low rates of selective miticides to reduce the pest mite population while leaving the predator population intact. Selective miticides that control spider mites but have little impact on the western predatory mite at lower-than-label to low-label rates include cyhexatin (Plictran), propargite (Omite), and fenbutatin-oxide (Vendex).

Insecticides applied to control navel orangeworm and other insect pests often kill the natural enemies of mites. Carbaryl and permethrin are especially destructive. Some native populations of western predatory mites in commercial orchards are resistant to organophosphate insecticides if the orchards have been frequently sprayed, but no native carbaryl- or permethrin-resistant populations are known.

Strains of western predatory mites that are resistant to some organophosphate insecticides and carbaryl have been developed in the laboratory, released experimentally in some orchards, and are now commercially available. Successful establishment of these predatory mites, or maintenance of native populations of western predatory mites, may reduce the number of applications and rate of miticides needed.

Special management practices are required in orchards to establish insecticide-resistant (carbaryl-organophosphate) predatory mites. Predatory mites need a period of at least 6 months, and in some cases as long as 2 years, after they are released to become established. Low population levels of pest mites must be maintained in the orchard to serve as a food source for predatory mites and ensure their survival. A ratio of at least 5 to 10 pest mites to 1 predatory mite is recommended (10:1 in early season, 5:1 in midsummer). If predatory mites are present but the predator:pest ratio is out of balance, use lower than label rates of a selective miticide to reduce the pest mite population. Use of selective miticides at rates lower than normal will ensure that some pest mites remain and predatory mites do not starve.

Carefully monitor predatory mites and spider mites during the period of establishment to ensure that 1) an adequate level of the pest population is present to serve as a food source; 2) the predators have become established; and 3) the pest mite populations are not reaching damaging levels. To maintain insecticide resistance in the predatory mites, insecticides will need to be applied on an infrequent basis, but timing for these applications has not yet been determined. For additional information on managing populations of pest and predatory mites, see *Managing Mites in Almonds: An Integrated Approach,* listed under Insects and Mites in References.

Sixspotted thrips, *Scolothrips sexmaculatus,* is a mite predator frequently encountered in almond orchards with high spider mite populations, but it often does not move into the orchards until mite populations are high. Both adults and larvae prey on all stages of the pest mites and can rapidly reduce high mite populations. However, this usually occurs after mites have damaged the trees.

Spider mite destroyer, *Stethorus picipes,* is a small lady beetle that feeds on mites, particularly in unsprayed orchards. The larvae are dull brown or black and have a velvety appearance. Like the sixspotted thrips, however, it does not become numerous until mite populations are

The most dependable webspinning mite predator is the western predatory mite. This photo shows how the western predatory adult (top and bottom) differs in appearance from the webspinning mite (center).

Sixspotted thrips adults are characterized by three dark spots on each forewing. They are efficient mite predators but move into orchards only when mite populations are high, often after damage to the trees has already occurred.

SIX-SPOTTED THRIPS

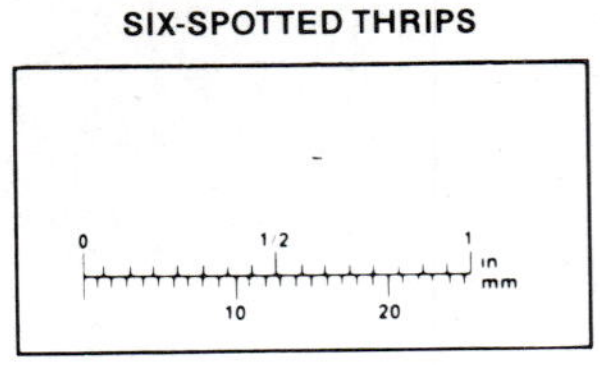

This small black beetle is an adult spider mite destroyer (Stethorus). It preys on mites but becomes numerous only when mite populations are heavy.

The larva of the spider mite destroyer also preys on mites.

STETHORUS

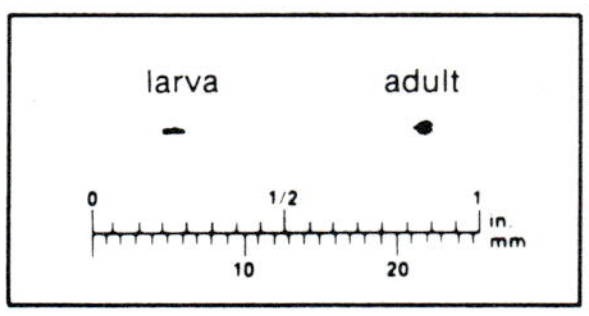

LACEWING

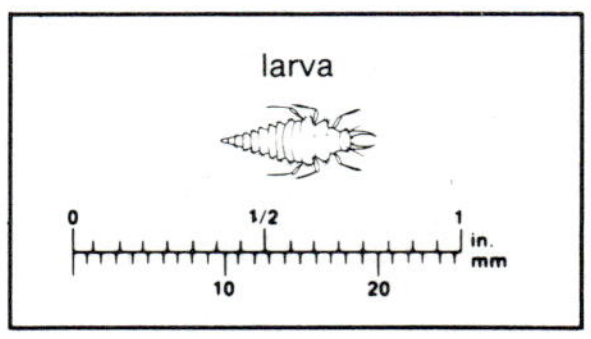

This adult green lacewing is devouring a mite.

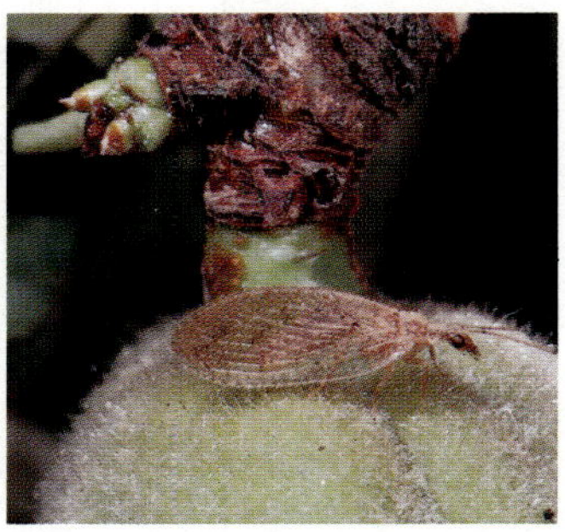

Brown lacewings feed on many species of pest mites.

Cecidomyiid larva feeding on mite eggs.

The green lacewing larva, pictured here, looks similar to the brown lacewing larva.

heavy. It is an active feeder, though, and can reduce mite populations quickly when it is abundant.

Brown lacewings, *Hemerobius* spp., green lacewings, *Chrysopa* spp., and larvae of a cecidomyiid predator, *Feltiella* sp., also feed on all stages of spider mites.

Monitoring. Check orchards for mites once every 2 weeks from March to the first part of May. From May though August, monitor the entire orchard weekly and problem areas, such as along the road or water-stressed trees, every few days.

For general monitoring in the early part of the season, observe individual leaves in the lower interior portions of trees throughout the orchard for signs of mite feeding. Use a hand lens to examine leaves for mite eggs and mobile mites. As a rule of thumb, if there are more eggs than mobile mites, the pest mite population is increasing. If you observe leaf stippling and low populations of mites are present, check also for western predatory mites. These predators often control mites in the early season before the populations can build up. Miticides in the early part of the season are, therefore, seldom required.

If you are managing a pesticide-resistant predator population and pest mite populations increase with the onset of hot weather, use lower-than-label rates of a selective miticide to adjust predator-prey ratios without removing all food sources for predators.

Two techniques (described below) for sampling mites in almond orchards from mid- to late season are brush and count sampling and presence-absence sequential sampling. Brush and count sampling can be used earlier in the season than the presence-absence sequential scheme

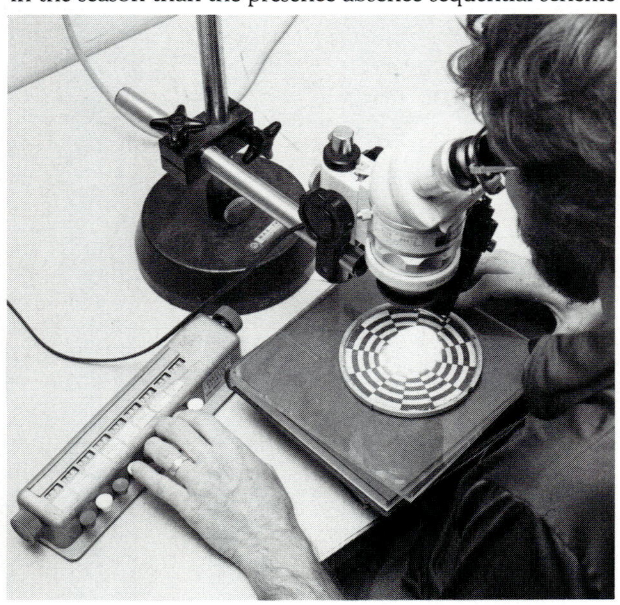

Figure 49. For brush and count sampling of mites, mites are counted under a microscope after they are brushed off the leaves onto a glass plate.

to assess population levels of webspinning mites, European red mites, and western predatory mites, but it is more time consuming than presence-absence sampling. The presence-absence sampling method, on the other hand, has not yet been developed to assess European red mite populations. You must be able to correctly identify active stages of both pest mites and western predatory mites to use these sampling methods. Photographs in this manual will help you to identify them.

Brush and Count Sampling of Mites

Begin sampling for mites with the brush and count method in early May. If webspinning mites are the only pest mites present in the orchard, switch to the presence-absence method in mid-May. If European red mites are consistently present in the majority of samples, you may want to use the brush and count method throughout the season.

To make the sample, a specified number of leaf samples are taken from selected trees, put in paper bags or closed containers, refrigerated, and taken to the laboratory for counting. There, the leaf samples are run through a mite-brushing machine, which removes the mites onto a glass plate. The glass plates are then placed over a paper grid and counted under a microscope (Figure 49). Specific sampling and treatment guidelines for brush and count sampling are available in *Managing Mites In*

WEBSPINNING MITES IN ALMOND TREES: A SAMPLING FORM

Date: ____________________

Orchard: ____________________

				PREDATORS PRESENT		PREDATORS ABSENT		
Tree number	Total no. leaves sampled	Number leaves with mites	Total leaves with mites	Don't treat if accum. total less than or equal to —	Treat if accum. total more than or equal to —	Don't treat if accum. total less than or equal to —	Treat it accum total. more than or equal to —	Number of leaves with western predatory mite and/or sixspotted thrips
1	15	____	____					____
2	30	____	____					____
3	45	____	____					____
4	60	____	____					____
5	75	____	____	27	40	12	24	____
6	90	____	____	33	48	15	28	____
7	105	____	____	39	55	18	31	____
8	120	____	____	45	62	21	35	____
9	135	____	____	51	69	23	39	____
10	150	____	____	57	76	26	43	____
11	165	____	____	63	83	29	46	____
12	180	____	____	70	90	32	50	____
13	195	____	____	76	97	35	54	____
14	210	____	____	82	104	38	57	____
15	225	____	____	88	111	41	61	____
16	240	____	____	94	118	45	65	____
17	255	____	____	101	125	48	68	____
18	270	____	____	107	132	51	72	____
19	285	____	____	113	139	54	75	____
20	300	____	____	119	146	57	79	____

STOP SAMPLING

Figure 50. Use this sampling form for presence-absence sequential sampling of webspinning mites.

Almonds: An Integrated Approach, listed under Insects and Mites in References.

Presence-Absence Sequential Sampling for Webspinning Mites

Presence-absence sequential sampling is presently applicable to webspinning mites only. Begin sampling in mid-May, and continue to sample through August to determine if a miticide spray is needed. To use this method, you do not need to count the numbers of mites per leaf; you need only to determine if they are present or absent.

1. Before July 1, you only need to monitor hot spots—that is, areas that usually develop mites first. These include roadside trees and trees in dry or dusty areas or areas with sandy streaks. Once a threshold is reached, sample the remainder of the orchard to determine if a spot treatment is sufficient or the entire orchard needs to be treated. After July 1, monitor the entire orchard by dividing it into sampling areas. A sampling area is the minimum portion of the orchard to which you're willing to make a treatment.

2. Within each hot spot or sampling area, randomly select trees to sample. Do not choose trees that are obviously different from others in the area, such as replants.

3. As you walk around the tree, randomly select leaves both inside and outside the tree canopy from all four quadrants of the tree. In sprinkled orchards, pick leaves from above the sprinkler line.

4. Sample a minimum of five trees within each sampling area. Select 15 leaves from each tree, and use a hand lens to examine both upper and lower leaf surfaces carefully for the presence or absence of webspinning mites and eggs, as well as predators such as western predatory mite and sixspotted thrips. Use a sampling form such as the one in Figure 50 to record your results. Don't count the mites; just record the number of leaves with webspinning mites or eggs in the first column of the sampling form and the number of leaves containing mite predators in the last column. Add together the total number of leaves with webspinning mites for all trees sampled and record this number in the second column. If western predatory mites or sixspotted thrips are present in your orchard, use the "Don't treat" and "Treat" figures given on the form under the heading "Predators Present." If no predators are present, however, use the treatment figures in the columns under the heading "Predators Absent." If the total number of leaves with mites falls between the "Treat" and "Don't treat" figures given on the form, continue to sample until a definite decision can be reached.

For example, if predators are present in the orchard and you find mites on 36 leaves out of 75 in the first five trees you sample, continue to sample until you either can make a treatment decision or you sample 20 trees. If, on the other hand, 15 out of 75 leaves in the first five trees contain mites, don't treat your orchard.

5. If a decision to treat is reached, consider the predator population. If the number of leaves with western predatory mites or sixspotted thrips is about equal to the number of leaves with webspinning mites, delay treatment for a few days; the predator population could cause the pest mite population to decline. If you do treat and predators are present in any abundance, use low rates of miticides to reduce the webspinning mite populations without eliminating them as food for the predatory mites. (If you are combining the miticide with a carbaryl hull split spray,

Eggs of the European red mite are red and have a characteristic stalk.

The adult female European red mite is dark red and has six to eight white spots at the base of hairs on the back.

European red mites cause leaf stippling. Dark spots on the surface of these leaves are adult European red mites.

however, and the predators in your orchard are not resistant to carbaryl, use the full miticide rate.)

6. If after sampling 20 trees you could not reach a decision, sample again 2 to 3 days later. If you did reach a decision, based on sampling, to not treat at the present, sample again in a week.

European Red Mite
Panonychus ulmi

European red mites can be found on nuts, pome and stone fruits, berries, and some ornamental shrubs. High populations may occur in almond-growing regions from Kern County in the south to northern portions of the Sacramento Valley.

Description and Seasonal Development

European red mites overwinter as eggs. Look for eggs in roughened bark at bases of buds and spurs on smaller branches and twigs, or in wounds. They are globular and red with a slender stalk (stipe) arising from the top center and many grooves extending from top to bottom. During the growing season, eggs are laid on the leaves. Overwintering eggs are laid as early as August and, depending upon the weather, through October and November.

Eggs hatch when the trees bloom, and immature mites migrate to leaf buds and feed on leaf juices. Three instars occur before the adult stage. Immature mites are bright red, except just after molting when they appear bright green. The green color turns to red after mites resume feeding. Adults are dark red and have six to eight white spots at the base of hairs on the back. These white spots distinguish European red mite from citrus red mite.

Depending upon temperature, European red mites can have five to ten generations per year. Low spring temperatures can keep high populations from developing. If these mites build up in an orchard during the spring, the population may survive the hot, dry conditions of summer and become generally evenly distributed throughout the orchard.

Damage

Feeding by European red mites causes leaf stippling. Heavy or prolonged feeding will eventually cause leaves to turn pale and appear bronzed and burned at the tip and margins; in severe cases it can even cause leaf drop. European red mite is not as damaging initially as webspinning mites, but repeated, heavy infestations may reduce yields. For example, heavy infestations for 2 years in a row on walnut trees reduced yield the third year by 40%, and the trees did not completely recover the first year when subsequently protected. Damaging effects have also been recorded on apple and plum trees, but the effect of European red mite on almond trees has not been measured.

European red mites are found mostly on the upper surface of leaves. They spin little or no webbing. As the population increases, the mites will colonize both surfaces of the leaf. Low numbers of European red mites, about two to three per leaf, early in the season are considered beneficial as a food source for the western predatory mite so that this predator can build up to numbers sufficient to control webspinning mites later in the season. In the process, these and other predators significantly impact the European red mite population.

Management Guidelines

A dormant spray of oil is fundamental to European red mite management. A properly applied oil spray during the dormant period that includes a suitable organophosphate will also control peach twig borer, San Jose Scale, and brown mite. In-season sprays for European red mite are usually not necessary if oil is applied with the dormant spray. Even if it does not provide complete control against European red mite, a dormant spray will often keep populations at low levels.

Biological Control. Brown lacewing, *Hemerobius* sp., is an effective European red mite predator, but its occurrence is unreliable and not easy to predict. Western predatory mites feed on immature and adult stages of European red mite but are unable to break through the egg shell. Therefore, they are not as effective against European red mites as they are against webspinning mites.

Monitoring and Control. The best way to control European red mite is with a dormant spray. Check orchards for European red mite once a week during the growing season, however, as part of your regular monitoring program. You can check leaves for European red mites at the same time you are looking for webspinning mites. No treatment guidelines are available for European red mite. European red mites are not as damaging as webspinning mites, however, and almond trees can tolerate a greater number of European red mites feeding per leaf than webspinning mites before economic loss occurs.

European red mites must be monitored more intensively in the spring than webspinning mites. Use a hand lens to examine leaves for mites and eggs. Also check the degree of leaf stippling that can be directly attributed to European red mite. European red mites may also be monitored with the brush and count method described in *Managing Mites in Almonds: An Integrated Approach*, listed under Insects and Mites in References.

Brown mites overwinter as red eggs that are laid in masses on twigs or under bark scales. Brown mite eggs do not have a stalk arising from the top. Adult brown mites are dark brown, and their first pair of legs are longer than the other three pairs.

Brown mite feeding causes leaves to turn whitish grey.

The adult citrus red mite closely resembles the European red mite but does not have white spots at the base of hairs on the back.

Peach silver mites are extremely small, light-colored mites.

Do not treat low European red mite populations because they help maintain predators for later control of webspinning mites. If extensive stippling is found in an area but predators are present, delay treatment a few days to see if the predators contain the mite population. If high mite populations persist, treat the area of infestation.

In-season insecticide sprays may disrupt biological control of European red mite and flareups may occur, so be sure to monitor carefully after insecticide treatments.

Brown Mite
Bryobia rubrioculus

Brown mite is a pest of stone and pome fruits throughout the world. It is the largest of the mites that feed on almonds and the first to appear in spring. If not kept under control with a dormant spray, brown mite can cause damage in the northern San Joaquin Valley and the Sacramento Valley.

Description and Seasonal Development

Like European red mite, brown mites overwinter in the egg stage. Eggs are red and spherical but do not have a stalk arising from the tops. They are laid in masses under bark scales and at the base of buds, and hatch at the same time the leaf and flower buds open, usually 1 to 2 weeks before European red mite eggs hatch. Hatch is completed before full bloom.

Newly hatched brown mites have six legs and are bright red. After they molt for the first time, they turn brown, have eight legs, and resemble the adult but are smaller. Immature mites feed mainly on underleaf surfaces.

Adult brown mites are dark reddish brown, and the first pair of legs is longer than the other three pairs. During the warmest part of the day, brown mites rest and reproduce in woody parts of the tree. When temperatures cool off, they move onto the leaves and feed on both upper and lower leaf surfaces. Frequently, brown mites only infest a few trees within the orchard.

Adult females live 2 to 3 weeks and reproduce without mating; there are no males. Brown mites are not active during hotter periods of the summer. Eggs laid during hot weather will not hatch until the following spring, and populations generally decline in mid-summer. Usually two or three generations of brown mite occur between February and early June.

Damage

Feeding by brown mites causes whitish gray spots to appear on leaves. Brown mites do not produce webbing and rarely cause leaf drop. Affected leaves first become mottled, then bleached but do not turn bronze at the tips as do leaves damaged by European red mites. Badly damaged foliage is smaller than normal and covered with minute flecks of dried feces.

Management Guidelines

Generally brown mites are effectively controlled by a dormant spray. Occasionally they are a problem in the Sacramento Valley, however, when there is a cool spring and inadequate amounts of oil were used in the dormant spray. Brown mites have the same predators as European red mites, and moderate numbers of brown mites in the spring can be beneficial in providing a food supply for these predators. In-season sprays for brown mites are generally not needed.

Citrus Red Mite
Panonychus citri

Citrus red mites are pests of almonds mainly in the southern portion of the Central Valley where citrus is grown. These mites invade almond orchards each spring when they are blown in from citrus groves. All stages of the citrus red mite closely resemble European red mites, but citrus red mites do not have white spots at the base of the hairs on the back.

Citrus red mites may occur in areas that also have persistent infestations of webspinning mites. If the citrus red mite population is not localized and occurs consistently throughout the sampling area, sample the orchard using the brush and count method described in *Managing Mites in Almonds: An Integrated Approach,* listed under Insects and Mites under References.

Peach Silver Mite
Aculus cornutus

Peach silver mites are sometimes found on almond foliage in the Central Valley. Although these mites may occur in large numbers, they are not known to be damaging. In fact, peach silver mites are often considered a beneficial species because they serve as a food source for mite predators in the early season.

These leaf feeders are extremely small and white. Because peach silver mites are much smaller than other mites on almonds, a microscope may be needed to see them clearly. Feeding by peach silver mites causes tiny, chlorotic spots that give the leaf a silvery appearance, especially along the midvein of the upper surface of the leaf. Symptoms resemble thrips or leafhopper damage. No treatment is necessary.

Pavement Ant
Tetrameorium caespitum
Southern Fire Ant
Solenopsis xyloni

Pavement ants and southern fire ants can cause economic damage to almonds when the nuts are drying on the ground after harvest. In the past this damage was not significant; however, recent increased planting of almonds in the southern San Joaquin Valley on previously nonirrigated soils and changes in orchard floor and

Ant hills in almond orchards often appear as tiny mounds or small patches of loose soil.

The pavement ant is brown and covered with coarse hairs.

The southern fire ant has an amber head and thorax with a black abdomen.

Ants feed on almonds by hollowing out the nut meats. Eventually all that is left is parts of the skin, as seen in the nut in the top of the photo.

water management practices may be leading to increased ant populations and ant damage in almond orchards. These practices include change to nontillage orchard management and greater reliance on cultural methods such as early harvest rather than insecticides for control of navel orangeworm.

Description and Development

Ant hills in almonds orchards often appear as tiny mounds or small patches of loose soil. The ant colony consists of one to several queens and thousands of workers. Queens lay eggs and do not leave the nest. Workers care for queens and forage for food. At certain times of the year, winged males and winged females (or queens) are produced. They leave the nest in swarms and disperse to new locations. Males die shortly after mating, while females locate a new nest site, tear off their wings, excavate small caves in the soil, and lay eggs. The queen feeds and cares for the young maggotlike larvae in her cave until the larvae are full grown. From that point on the larvae take care of the queen, whose sole function is to lay eggs.

Although other ant species have been found in orchards and have caused occasional damage, the pavement ant and the southern fire ant are the only ones that have been associated with major damage to almonds. Black harvester ants (*Veromesor andrei*) will feed on almonds, but generally are not very damaging because they are too large to enter the nuts.

The pavement ant worker is light to dark brown and covered with coarse hairs. Workers travel in single file in search of food and usually nest in sandy or loamy soils where the soil temperature is 58° to 82°F (14° to 28°C).

The southern fire ant has an amber head and thorax with a black abdomen. This ant has a painful sting that can cause visible swelling. It is found throughout the San Joaquin and Sacramento valleys and prefers soil temperatures of 70° to 95°F (21° to 35°C).

Damage

The southern fire ant has a wider distribution and causes more damage in California almond orchards than does the pavement ant. Ants feed on a wide variety of plant material in the orchards. Ants usually do not feed on almonds until they are on the ground, and damage increases in proportion to the length of time the nuts are there. Damage by both species of ants first appears as a scraping or peeling of the pellicle, or skin of the kernel. Eventually the meat is completely hollowed out leaving only parts of the skin. Unlike navel orangeworm damage, there is no frass or webbing, although white, sawdustlike material from the ant's chewing may remain. In some orchards, ant-damaged nuts may account for the major portion of the nuts rejected at the processor.

ANTS

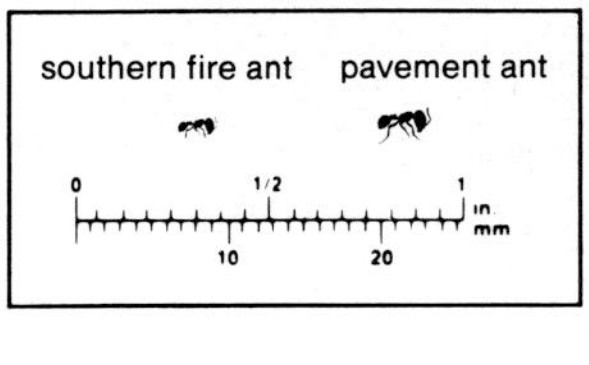

Management Guidelines

If you think ant damage is a problem in your orchard, survey the orchard floor for colonies in the spring or shortly after an irrigation. Ant hills are easiest to see 2 to 3 days after an irrigation because the soil on the top of the mound is freshly turned. Later in the season, especially when the orchard is being dried for harvest, ant hills are difficult to see. Also, when the soil cracks, ants may travel in the cracks rather than on the soil's surface and are harder to see.

Before harvest, reexamine infested areas more critically; examine the area from mid-alley to mid-alley along ten trees and count the number of active ant colonies. For each block evaluate a minimum of five such areas. It may be difficult to determine where one colony begins and another ends. Large colonies may have several entrance holes separated by more than 3 feet. Active colony entrances more than 3 feet apart, however, should be counted separately. Ant activity appears to peak in the morning and again just before sunset; these are the best times to identify active colony entrances.

Once you have determined the number of colonies per sampling area, check Table 10 to find the corresponding estimate of damage. Note that the damage increases with the number of days that nuts remain on the ground. If your colony counts indicate that some areas may have a level of damage that is not acceptable to you, apply a treatment to those areas.

Table 10. Percent Damage to Almonds Caused by Ants after Harvest.

No. of colonies per 5,000 square feet[1]	Days postharvest				
	4	7	10	14	21
15	0.9	1.6	2.1	3.1	4.9
45	1.4	2.3	3.2	4.7	7.0
185	2.0	3.6	5.0	7.0	11.1

1. The area from the middle of the alley on one side of ten trees, spaced 22 by 22 feet, to the middle of the alley on the other side is roughly equal to 5,000 square feet.

Hull split sprays for navel orangeworm may adequately suppress worker ants, but the colonies will eventually recover and persist from year to year. Ants are easily controlled with an application of a broadcast granular insecticide, but use these materials only when monitoring indicates a damaging number of colonies. Granular insecticides take about 2 weeks to destroy the ant colony. Apply granules early enough to get complete control before harvest and to comply with the preharvest restrictions—anytime from May through July. Irrigating after applying granular materials appears to not only increase the effectiveness of the material but to extend the effective period as well. Annual applications may not be required, but it is good practice to monitor your orchards each year to determine if a treatment is needed. Granular formulated insecticides applied directly to the ground have a minimal effect on nontarget beneficial insects and mites in the orchard.

Oriental fruit moths are small and gray to brownish white.

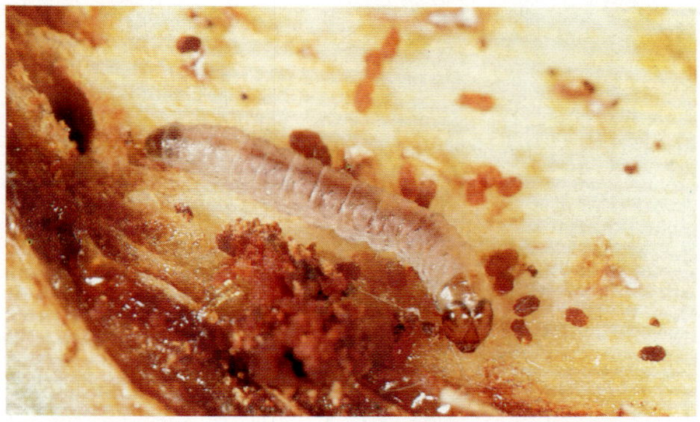

The oriental fruit moth larva is white or pink with a brown head.

Oriental Fruit Moth
Grapholitha molesta

Oriental fruit moth is an occasional pest of almonds. Early in the season, larvae bore into the tips of tender twigs and cause them to wilt and die. This damage can make it difficult to train first and second year trees. Although oriental fruit moth larvae mine deeper into a shoot than peach twig borer larvae, distinguishing oriental fruit moth damage from peach twig borer damage can be difficult, unless the larva is still in the twig. The oriental fruit moth larva is white or pink with a brown head; the peach twig borer larva is dark brown with a black head.

Oriental fruit moth larvae also may be found feeding between the hull and shell after hull split, when the hulls are still fairly green. In addition, the larva can enter the shell to feed on the kernel. Until recently, oriental fruit moth was not considered an economic pest of the nutmeats. During the 1983 growing season, however, oriental fruit moth infestations in nuts occurred in Fresno, Madera, and Butte counties. In other crops, oriental fruit moth larvae prefer mature fruit, but in almonds they have been found boring into nuts in late May to early June. At hull split, the larvae enter the nuts primarily through splitting sutures and at the point of stem attachment.

Like navel orangeworm, oriental fruit moth may feed in groups of several larvae within a single nut. An oriental fruit moth larva can be distinguished from navel orange-

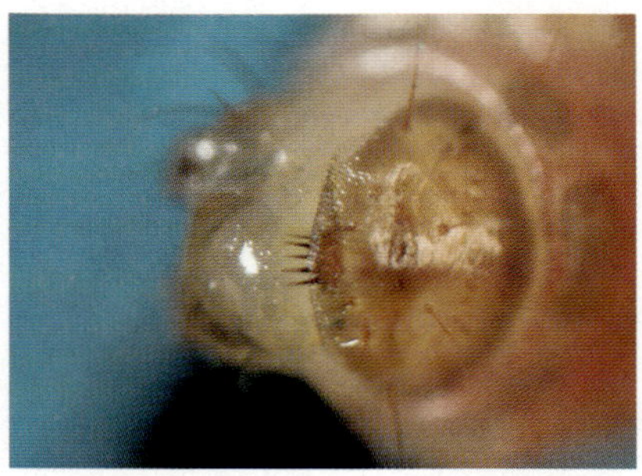

Oriental fruit moth larva can be distinguished from other larva, such as the navel orangeworm, by an anal comb.

ORIENTAL FRUIT MOTH

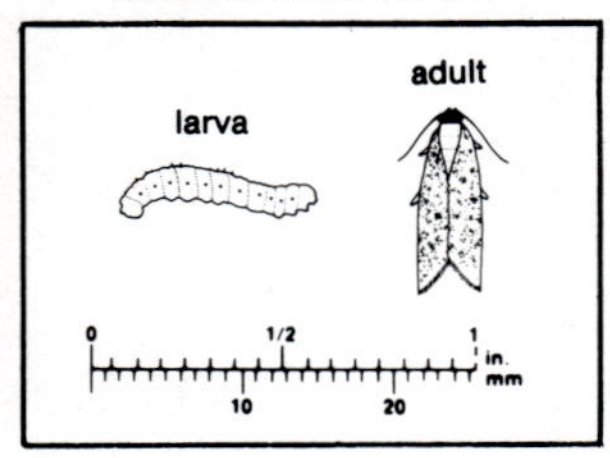

Nut damage caused by oriental fruit moth larvae is similar to that caused by peach twig borer larvae; they do not bore deeply into the nut meats and do not produce webbing.

Peachtree borer larvae feed in the crown area of trees.

When monitoring for peachtree borers, look for small accumulations of reddish brown frass on the lower tree trunk, such as on the tree here. The empty pupal case lying at the base of the tree indicates that a peachtree borer moth has recently emerged.

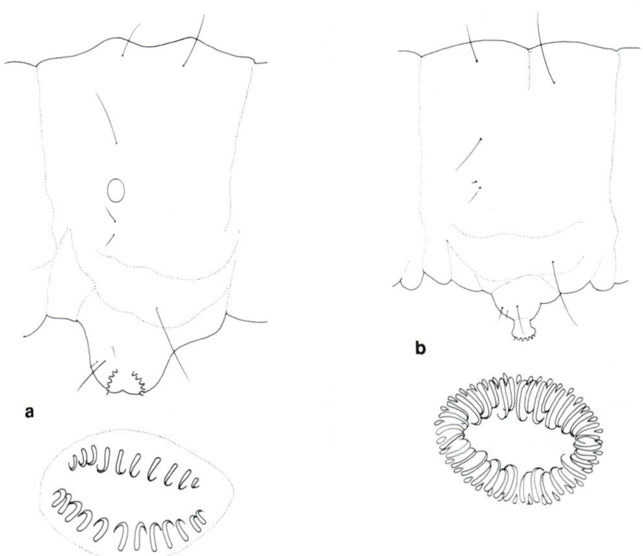

Figure 51. Upper diagram is the lateral view of fourth abdominal segment of (a) peachtree borer and (b) American plum borer. Note the size and shape of the proleg in relation to the abdominal segment. The lower diagram illustrates the crochet pattern that can be seen with a hand lens on the bottom of the proleg.

worm by the absence of the pair of crescent-shaped marks on the second segment behind the head and the presence of an anal comb. Also, the oriental fruit moth larva does not produce webbing, and it produces reddish brown frass in the hull, in contrast to the brown and white frass of the navel orangeworm. Damage to the kernel by oriental fruit moth resembles that caused by peach twig borer.

A method for monitoring this pest with pheromone traps, estimating when generations will occur, and ideal time for treatment using degree-days is outlined in *Monitoring Oriental Fruit Moth Development with Day Degrees* listed under Insects and Mites in References.

Peachtree Borer
Synanthedon exitiosa

Peachtree borer is a pest of tree crops in the Santa Clara Valley and in Contra Costa County. It is also becoming a problem in western San Joaquin, Stanislaus, and Merced counties.

Peachtree borers overwinter as larvae in the tree trunk. Larvae have light colored bodies and dark heads. Figure 51 illustrates how to distinguish peachtree borer larvae from American plum borer larvae, another pest that can be found boring in the crown area. Peachtree borers feed in the crown area and burrow up into the tree. In late spring they pupate near the entrance of their burrows or in the soil. Steel blue to black moths emerge in 20 to 30 days, and the adult flight period extends from May through September. Eggs are laid during the summer on the bark of tree trunks. When the eggs hatch, the young larvae emerge and tunnel into the tree at or slightly below ground level. One generation occurs per year.

Peachtree borers can girdle and kill a young tree. Older trees are sometimes attacked but can sustain the damage, unless there are many larvae or the tree is attacked several years in a row.

Management Guidelines

Almonds grown on peach rootstock are more susceptible to peachtree borer infestations. If peachtree borers are

PEACHTREE BORER

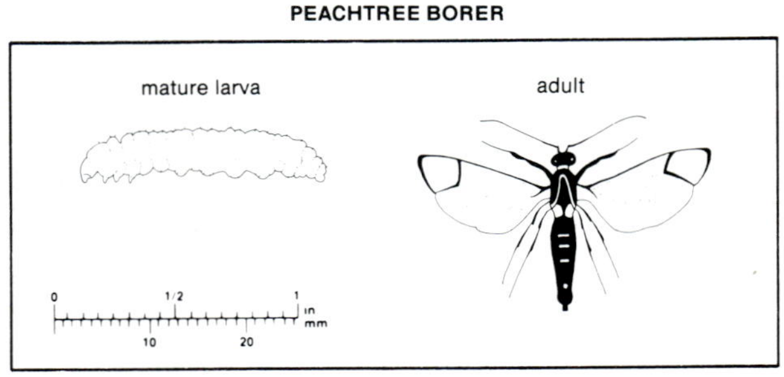

a problem in your area, weigh potential damage by this pest along with other factors when choosing rootstock for new trees.

Damage to trees can occur above or below the soil line. If you live in an area where peachtree borers are a problem, examine your trees yearly. Pheromone traps are available to monitor populations of peachtree borer moths and determine if they are in your orchard. Consult your farm advisor for more information on the use of these traps.

The best time to look for peachtree borer is in the fall. Look for small accumulations of frass on the lower tree trunk, especially on trees that appear weakened. After the first rains, sap will exude from the larval tunnels. Remove 6 to 8 inches of soil from around the trunk and peel back the bark to observe the burrows. If infestations are found, apply a chemical treatment the following season in mid-May and in mid-July to trunks of infested and adjacent trees to kill newly hatched larvae when they begin to bore into the trunks. If the infestation is extensive, treat the entire orchard. Remove soil from the base of the tree trunk and spray the trunk from the scaffold crotch to the ground so that some of the spray solution puddles and drenches the area below the soil surface.

American plum borer larvae can be found boring in the crown area of the tree.

American plum borer larvae also bore in the scaffold crotches of young trees.

Young trees damaged at the scaffold crotch by plum borer larvae frequently lose the weakened limb during windy periods.

AMERICAN PLUM BORER

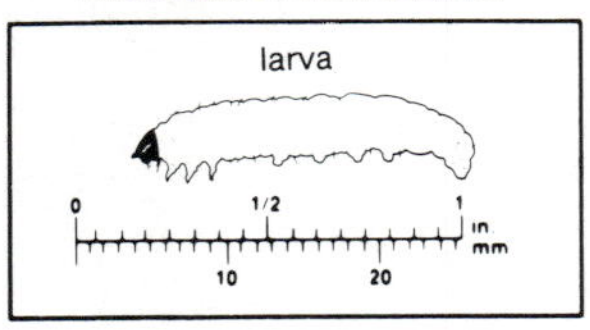

American Plum Borer
Euzophera semifuneralis

American plum borer is a sporadic pest in young almond orchards from Tehama County in the north to Merced County in the south. It occurs on all major almond cultivars, but first and second leaf Carmel and Price cultivars are attacked most frequently. It also infests other stone and pome fruits.

The plum borer has about three to four generations a year, and each generation takes about 4 to 6 weeks for the insect to complete development from egg to adult. The plum borer overwinters as a mature larva in a cocoon within the tree. The majority of moths emerge in April and May. Following mating, females lay opaque, white eggs near pruning wounds, in scaffold crotches of young trees, near graft unions of young trees that have been recently topworked, or on crown galls of older trees. Young larvae are active and have white bodies with large, dark brown heads. When mature, the larva is about 1 inch (25 mm) long; its body is dull white, pinkish, or dull green. Figure 51 illustrates how the American plum borer can be distinguished from the peachtree borer.

American plum borer larvae bore into the tree and tunnel in the cambium layer; reddish orange frass, webbing, and gum pockets indicate their presence. These larvae are most damaging when they tunnel in the scaffold crotches or graft unions of young trees. Vigorous trees usually heal over and continue to grow following a light infestation. Prolonged, heavy infestations, however, can weaken or girdle a scaffold, which then breaks off during windy periods, or when the tree has its first heavy crop. After the fifth year, trees are rarely attacked by plum borer larvae, possibly because the bark is thicker and tough enough to prevent penetration by this insect.

Monitor young orchards in spring and summer for plum borer larvae. Examine scaffold crotches for small piles of reddish orange frass, webbing, and gum pockets. If plum borer larvae are present, spray trees with a solution of paint, water, and insecticide. To make this solution, mix one part white interior latex paint to three or four parts water. The actual amount of water will be determined by the quality of paint used; if the paint is low quality, use three parts water. Add an insecticide at label rate to the paint mixture. The paint serves to lengthen the life of the insecticide.

Spray the tree from 1 foot above to 1 foot below the scaffold crotch about two to three times during the growing season. Make your first application from mid- to late April and subsequent applications at about 6-week intervals. This spray will repel plum borer invasions, as well as kill some of the larvae that are already present in the tree. Research is underway to obtain more information about this pest and the threat it poses to California almonds. For more information, consult your farm advisor.

Larva of the Pacific flatheaded borer feed in previously injured aboveground portions of the tree.

The adult Pacific flatheaded borer is often found in moth pheromone traps.

PACIFIC FLATHEADED BORER

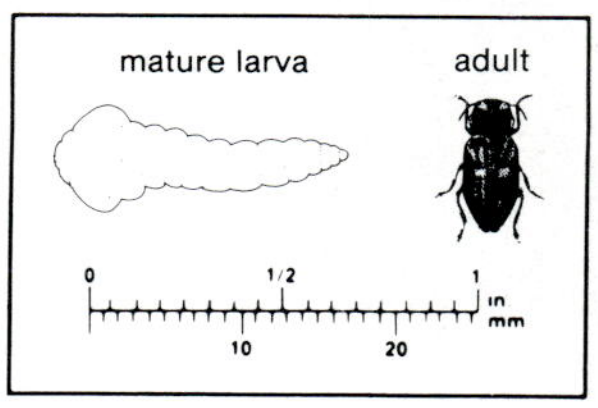

Adult female shothole borers are tiny black beetles.

SHOTHOLE BORER

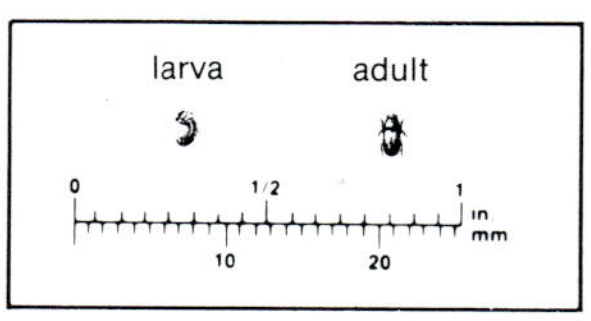

Shothole borers invade the bark of previously damaged trees and lay their eggs in tunnels that can be seen on this limb where the bark has been cut away. Externally all that can be seen is resin exuding from small holes.

Pacific Flatheaded Borer

Chrysobothris mali

Pacific flatheaded borers are attracted to diseased or injured limbs, such as those affected by sunburn, scale insects, bacterial canker, mallet wounds, or major pruning cuts. Unlike peachtree borers, which attack undamaged tree trunks, Pacific flatheaded borers attack above ground portions of the tree that have been previously injured. These beetles lay eggs in the injured area, and the larvae excavate large caverns just beneath the bark and bore tunnels deep into the tree's cambium tissues. Excavations are usually filled with finely powdered sawdust. A full grown larva is light colored, with a prominent, flat enlargement of the body just behind the head. The adult beetle is about 3/8 inch (10 mm) long with dark bronze bodies and have coppery spots on the wing covers.

Injury by this borer will cause sap to flow, and the affected area will appear as a wet spot on the bark. Later, these areas may crack and expose the mines. Feeding by Pacific flatheaded borers may cause a portion of the bark to die, or girdle and kill young trees.

Management Guidelines

Flatheaded borers often invade sunburned areas on the trunk of newly planted first year trees. Wrapping or painting the tree trunk above and 1 inch below the soil line with white, water-based paint or whitewash will protect the trunk from sunburn and flatheaded borer invasions.

In older trees the best way to avoid infestations is to keep your trees sound and vigorous. Prune out all badly infested wood, and burn or remove it from the orchard before the growing season starts. Spraying for this insect is not recommended.

Shothole Borer
Scolytus rugulosus

The shothole borer is a small beetle that usually attacks trees already weakened by root diseases, insufficient irrigation, chronic mite infestations, sunburned limbs, Pacific flatheaded borer, or the like. Adult females bore tiny holes in the bark and lay eggs in the cambium layer of the tree. When the eggs hatch, young larvae feed and excavate secondary galleries at right angles to the egg gallery. The outline of the gallery system resembles a centipede.

Normally, a number of shothole borer adults invade a tree at the same time. Healthy trees exude resin, which usually kills the insects. If the tree has injured or weakened areas, this resin buildup does not develop and the invasion is successful. Ultimately the larvae may girdle the tree, or tree part, and cause its death.

Management Guidelines

Shothole borers invade trees that have been previously damaged. The nature of this damage dictates the course of preventive action. Maintaining trees in a sound and vigorous condition, with sufficient fertilizers and water, will prevent attack by this beetle and keep uninfested tree limbs from damage. Often shothole borers occur in young trees that are already infested with American plum borer or Pacific flatheaded borer. If the tree is vigorous in all other respects, prune out the infested scaffold. Pruning can also be helpful in eliminating wounded areas in older trees infested with shothole borer. Severely infested trees should be removed. Burn or remove all infested wood from the orchard before the growing season starts. Do not leave pruned limbs or stumps (healthy or infested) near orchards (for example in woodpiles) as populations can emerge from these materials before they dry out, and beetles will then migrate into orchards. Spraying for this insect is not recommended.

Boxelder Bug
Leptocoris trivittatus

Boxelder bugs are a more serious problem in Sacramento Valley almond orchards than in the San Joaquin Valley. The bugs' dark, oval body has many fine red lines on its back. The red undersides of the wings can be easily seen when it is in flight. These insects can invade almond orchards anytime after blossoms fade through the end of May when their preferred hosts (maple, oak, elderberry, and boxelder) leaf out. Adults overwinter in grassy areas. Migration into almond orchards depends on how long the overwintering site remains suitable. In a cool, wet year, overwintering vegetation may remain succulent until the preferred hosts leaf out, and the bugs may not enter the almond orchards at all. On the other hand, mowing the overwintering area may trigger an invasion.

Boxelder bugs feed directly on young nuts. In the most severe cases, feeding in green nuts before shell hardening can cause the embryo to wither and abort. Larger nuts can sustain several attacks without withering, but the mature kernel will have small dark spots on its surface. Larger, damaged nuts gum internally, and during April some will drip gum in long, twisted threads. If you cut into the damaged nuts, the string of gum will lead down to a bump on the shell, where the bug's mouthparts entered the nut. A small, dark depression will be evident on the nutmeat just under this bump.

Once the shell hardens in late May or early June, the bug can no longer damage the crop because its mouthparts cannot pierce the shell.

Management Guidelines

No registered chemical treatments exist for this insect in almonds. It does not occur yearly in almonds, and usually by the time the damage is noticed, it has left the orchard.

Leaffooted Bug
Leptoglossus clypealis

The leaffooted bug can be recognized by the tiny leaflike enlargments on its hind legs. The narrow, brown body

Adult and eggs of the boxelder bug.

BOXELDER BUG

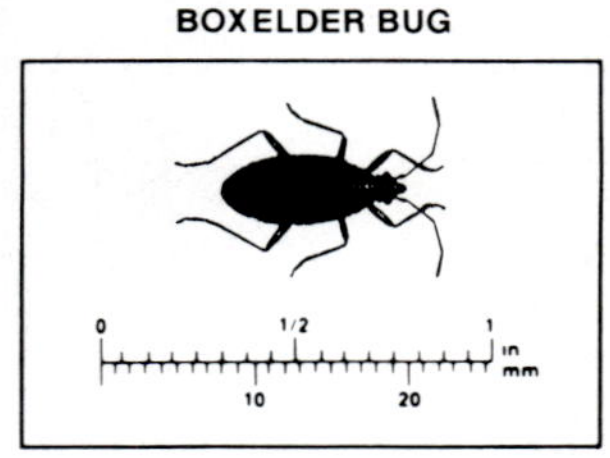

A leaffooted plant bug has tiny, leaflike enlargements on its hind legs and a yellow zigzag line across its back.

TOM EICHLIN

The carob moth is silver with black wavy patterns on its forewings.

The consperse stink bug is shield shaped and grayish brown with small black spots.

RICHARD E. RICE

Nuts damaged by bugs have small, dark spots on the surface and may be distorted in shape.

MAX BADGLEY

The carob moth pupa (top) has a raised dark ridge in the middle of the thorax and two short dark spines on each abdominal segment. These features are not present on navel orangeworm pupae (bottom).

has a yellow zigzag line across its flattened back. Found mostly in the Fresno area, it occasionally invades orchards in the early spring and causes damage similar to the boxelder bug. This insect is an infrequent pest of almonds, and no controls are recommended.

Consperse Stink Bug

Euchistus conspersus

The consperse stink bug is shield-shaped and grayish brown with small black spots. Like the boxelder and leaffooted bugs, it occasionally invades almond orchards and causes gumming and spotting of the kernel. High populations occur sporadically in the San Joaquin Valley, usually in late May or early June at intervals of 7 to 10 years. Control measures are not recommended.

Carob Moth

Spectrobates (=Ectomyelois) ceratoniae

The carob moth is closely related to the navel orangeworm. Although this species was only recently discovered in California, a subsequent survey indicates it is widely distributed south of the Tehachapi Mountain Range. It occurs in the Mediterranean region, where it is a pest of almonds and other crops such as dates, citrus, carob, fig, quince, and pomegranates. The carob moth not only looks like the navel orangeworm, it has a similar life cycle and causes the same type of damage to the almond crop. It lays eggs directly on mummies or hull split fruit, where both larvae and pupae can be found.

It is easiest to distinguish the carob moth from the navel orangeworm when it is in the pupal stage. The tan colored carob moth pupa has a raised dark ridge in the middle of the thorax that extends from the head to abdomen. In addition, each abdominal segment has two short, dark spines, and two small anal hooks protrude from the posterior end.

To distinguish carob moth larvae from navel orangeworm larvae, look for the characteristics illustrated in Figure 52. Because the carob moth is not currently found in the Central Valley, notify your agricultural commissioner if you think it is present in your orchard.

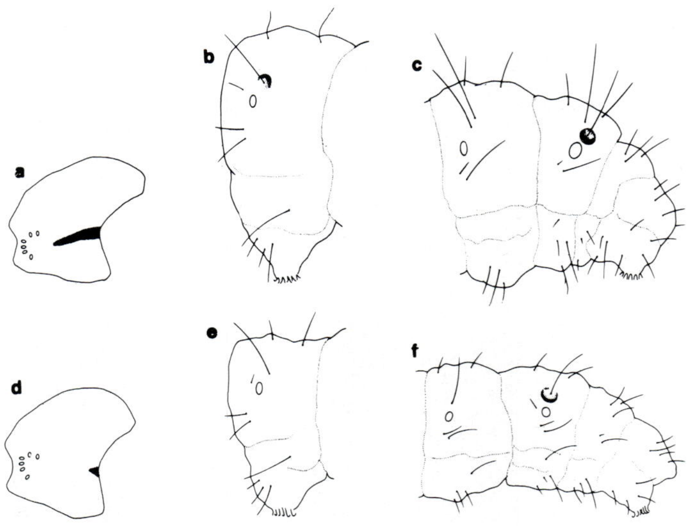

Figure 52. Because distinguishing characteristics can be variable, look at more than one feature to distinguish carob moth from navel orangeworm larvae. Carob moth larvae have a darkened band on the side of the head capsule (a); navel orangeworm larvae do not have this band (d). The hair just above the spiracle on abdominal segments one through seven has a pigmented area above its base in carob moth larvae (b), but not in navel orangeworm larvae (e). On the eighth abdominal segment, the pigmented area at the base of the hair usually forms a full circle in carob moth larvae (c), but only a horseshoe-shaped crescent in navel orangeworm larvae (f). (From a drawing by Dr. Thomas Eichlin, CDFA.)

Fruittree Leafroller
Archips argyrospilus

The fruittree leafroller is a green caterpillar with a black head and plate just behind the head. The head color changes to brown and the plate to dull green as the larvae mature. Caterpillars feed on tender new leaves that they roll and tie together with silken threads. When disturbed, they wiggle vigorously. Fruittree leafrollers may also enter young almond fruits and devour the kernel. By May, damaged nuts are dry and collapsed with large slotlike holes into the empty kernel chamber. The number of nuts attacked is usually insignificant; this insect rarely requires control measures.

Western Tent Caterpillar
Malacosoma californicum

Western tent caterpillars are hairy and dull yellow with rows of blue and orange spots along their backs. In spring they build silken tents in trees where they feed on foliage. Heavily infested trees may be defoliated, but this insect occurs infrequently in almonds and is rarely treated.

RICHARD E. RICE

The fruittree leafroller larva is a green caterpillar with a black head. It occasionally feeds in young nuts.

R.A. VAN STEENWYK

The western tent caterpillar has yellowish hairs covering its dark body.

Diseased trees, such as this one with Phytophthora crown and root rot, frequently exhibit similar overall symptoms: light green foliage, weak growth flushes, leaf drop, and twig dieback.

Diseases

Almond diseases are caused by a number of microorganisms and environmental stresses. Most diseases do not affect all almond cultivars to the same degree or occur in all almond-producing districts; also, little is known about the susceptibility of recently released cultivars. Many diseases, especially those affecting flowers, nuts, and leaves, occur only at a particular time of the year; other infections are present in the tree and produce symptoms or weaken the tree throughout the year.

Severity of a disease is influenced by virulence of the pathogen, susceptibility and maturity of the tree, and such environmental variables as temperature, rain, humidity, and soil moisture and texture. The impact of the pathogen on the tree also depends upon where infection occurs on the tree (Table 11). For example, trees with crown and root rots usually exhibit similar above ground symptoms. Damage to crown or roots disrupts uptake and transport of water and nutrients to and from roots. Affected trees variously show poor shoot growth, small chlorotic leaves, premature defoliation, decreased productivity, dieback of terminal shoots, and subsequent death.

Trunk and branch cankers damage bark and rapidly growing cambium tissues just beneath the bark. Cambium cells produce food-conducting tissue, or phloem, in the inner bark and water- and mineral-conducting tissue, or xylem, in the wood. Collapse of the cambium layer prevents production of new cells for these conducting tissues. Symptoms—including reduced growth and wilting or dying leaves—reflect partial or complete cutoff of food and water, depending on how much tissue is affected. If a canker kills the bark of the tree as deep as the cambium around the entire circumference of a trunk or branch, the portion of the tree above the canker usually will die within a few months; this symptom is called girdling.

Vascular pathogens invade and plug the xylem vessels, halting movement of water and nutrients up from roots. In the case of Verticillium wilt, leaf and branch symptoms are similar to those caused by crown and root rots, while symptoms associated with almond leaf scorch are similar to those caused by an excess or deficiency of certain mineral elements.

Table 11. Diseases of Almonds and the Location of Infection.

Crown or roots	Trunk and branch	Vascular system	Branch, foliage and/or fruit
Phytophthora crown and root rot	Ceratocystis canker	Verticillium wilt	brown rot blossom and twig blight
crown gall	bacterial canker	almond leaf scorch	bacterial blast
oak root fungus	Phytophthora canker		shot hole
wood rotting fungi	foamy canker		corky spot
	band canker		scab
			leaf blight
			green fruit rot
			yellow bud mosaic
			hull rot

Some pathogens cause infections that are limited to branches, foliage, and fruit. They produce localized lesions or larger infections such as shoot strikes. Infections frequently occur after rains or sprinkler irrigation when water gets into the tree's canopy, spreading the pathogen and creating conditions favorable for infection. The tree is also susceptible to infection by many pathogens during bloom because flowers provide a good entry point.

Major almond diseases include Phytophthora crown and root rot, Ceratocystis canker, bacterial blast and canker, and shot hole. Phytophthora crown and root rot is a major problem because it can kill the tree, and few controls, aside from careful water management and use of resistant rootstocks, are available. Serious losses can occur during extended wet periods caused by heavy rainfall or over-irrigation. Ceratocystis canker is associated with bark injuries created during harvesting operations. The canker eventually kills the affected limb of the tree, or the entire tree if it is located on the main trunk.

Bacterial blast and canker can be serious in young orchards growing in coarse-textured soils. Bacterial canker frequently kills 2- to 5-year-old trees, and no controls are available for either the blast or canker phase of the disease. Shot hole, a disease of leaves and fruit, can defoliate the tree and cause crop reduction in many almond-growing regions if not controlled annually with fungicides.

Monitoring and Diagnosis of Almond Diseases

Familiarize yourself with symptoms of common almond diseases, and watch for signs of tree stress as part of regular orchard monitoring. Watch for weak trees and irregular growth patterns. Check low-lying sites and areas with fine-textured or shallow soil; trees growing in these locations tend to have a higher incidence of diseases. Monitor more frequently during and after periods of rainfall. To diagnose a problem, examine the trunk for gumming, discoloration, or sunken areas. Also examine roots, branches, leaves and fruit. You may have to scrape off bark, dig up some feeder roots, or remove soil from the crown or main lateral roots. A shovel, hatchet, chisel, knife, and hand lens are useful tools for field inspection. The text and illustations of specific diseases in this manual will also help you to identify disease symptoms.

Time of year influences what symptoms you may see; fungal and bacterial diseases of fruit and leaves cause most damage during rainy or foggy periods, or in sprinkler-irrigated orchards where the water hits the foliage and fruit. Although they usually infect trees in fall, pathogens attacking roots and trunks may not produce symptoms until the following spring, when the tree begins to actively grow but is unable to get sufficient quantities of water and nutrients from the soil.

Certain cultivars are prone to disorders that are not caused by pathogens and may not be evident until long after the trees are in production, or unless the trees are grown in certain environments. Less commonly, a mutation may occur in part of a tree, affecting fruit shape, color, or growth habit of leaves and twigs.

Many diseases ultimately have the same effect: a declining tree with light green foliage, weak growth flushes, leaf drop, twig dieback, and reduced crop. Primary symptoms may appear on roots and trunk, fruit, leaves, and twigs. They can affect overall growth habit, size, and crop yield. These symptoms help identify a disorder. In this manual, individual diseases are organized according to where the pathogen infects the tree. Keep in mind that other pests (for example, nematodes), nutrient deficiencies, poor water management practices, or the environment may also be the cause of decline.

Compare your field observations with the descriptions and photographs in this section. A combination of symptoms usually will identify a disease. Some diseases, however, are difficult to identify by field symptoms, and positive identification will require laboratory diagnosis. If you find symptoms not described here, see References for publications on diseases, or consult your farm advisor.

Keep accurate records of your observations, and note infected trees, or those suspected of being diseased, on a map of your orchard. Positive diagnosis, especially of slowly developing viral diseases, is often not possible at first inspection. Repeated monitoring and mapping will help you confirm your diagnosis, follow development of a disease, and make proper management decisions.

Prevention and Management

The pathogen, the almond tree host, and the environment must all be considered in a disease management program. The pathogen is usually difficult to control; thus, emphasis lies on manipulating the tree and its environment to prevent disease outbreaks or at least lessen the impact of a disease. Prevention is especially important in disease management because many pathogens establish themselves slowly over years. Once disease symptoms are apparent, treatment options may be limited.

Preventive measures include preplant soil preparation, resistant rootstocks and cultivars when available, careful water management, sanitation, and fungicidal sprays. In addition, adequate fertilization, pruning, and control of other pests (insect and mites, weeds, vertebrates, and nematodes) also help maintain vigorous trees that are more resistant to infection than stressed, weakened trees.

Disease management in almonds starts before planting the trees. Fewer problems occur if the site has deep, level soil with good drainage. Low areas and finely textured and shallow soils have the potential for water logging. Where hardpans obstruct water penetration, ripping or backhoeing completely through the hardpan before planting helps improve water penetration.

Good water management is essential for maintaining healthy roots and vigorous trees throughout the life of the orchard. A carefully designed irrigation system and irrigations scheduled according to physical and climatic conditions of the site and needs of the trees will also contribute to disease management (see pages 24 to 27). As a general rule, do not irrigate during bloom, and use low angle sprinklers in spinkler-irrigated orchards so that leaves and nuts do not get wet. In poorly drained soils, use sprinklers or drip irrigation systems. Controlling weeds, especially around the base of trees where moisture favors development of crown rot, can also help prevent disease problems. Also, during periods of water stress, competition from weeds can further weaken trees making them more susceptible to certain diseases.

To prevent or reduce the impact of diseases, select resistant rootstocks and cultivars (if available), fumigate planting sites where almonds or related fruit trees have been pulled out, and pretreat the rootstock to protect against crown gall. Tolerant cultivars are available for some common diseases, although no cultivar-rootstock combination is resistant to all important diseases and environmental stresses. The choice, therefore, depends on prevailing local disease problems, on soil conditions and climate, and on cultural and marketing considerations. For more information on choosing rootstocks and cultivars, applying pesticides, irrigating, and preplant fumigating, see the chapter, Managing Pests in Almonds.

CROWN AND ROOT ROTS

Crown and root rot diseases disrupt uptake and transport of water and nutrients to and from roots. Above ground symptoms of affected trees are poor shoot growth, small chlorotic leaves, premature defoliation, decreased productivity, dieback of terminal shoots, and subsequent death. Many of these symptoms could be caused by factors other than crown and root rot pathogens. If your trees exhibit the above symptoms, remove soil from around the crown and main lateral roots to look for symptoms outlined in the following crown and root diseases.

Phytophthora Crown and Root Rot
Phytophthora spp.

Phytophthora crown and root rot kills more almond, apple, cherry, nectarine, peach, and walnut trees in California than any other disease. Crown and root rot is caused by several species of *Phytophthora*, a group of water mold fungi. These fungi are widespread and found in many orchards. They do not require wounds to penetrate the bark: if a favorable environment exists, they infect directly. *Phytophthora* infections occur most frequently throughout fall, winter, or early spring during periods of cool soil temperatures and abundant or excessive water.

Symptoms and Damage. At least eleven species of *Phytophthora* infect rootstocks on which almond trees are grown, and each is slightly different from the others in respect to the time of year it is most active, the portion of the tree it primarily attacks, and its temperature and moisture requirements. The pathogen may invade and infect roots, crown, trunks, or branches, or a combination of these, causing decay of invaded roots and of bark on the crown, trunk, or branches. Primary symptoms include root rot, crown rot, and trunk or branch cankers. A secondary, but more obvious symptom, is overall tree decline.

A tree infected with *Phytophthora* can either undergo a period of slow decline that may last a few years, or it can suddenly collapse and die in spring or early summer with the advent of warm weather. Severity and rate of disease development depends in part upon the combination of rootstock and species of *Phytophthora* involved.

Slow decline is a result of root infection by certain *Phytophthora* species that primarily cause feeder roots to die. Root rot is generally limited to secondary roots, but some *Phytophthora* species cause massive infection and decay of the entire root system that results in a rapid decline and death of the tree. In the early stages of root infection, diseased trees can be hard to distinguish from healthy ones. As the root rot progresses, infected trees develop open canopies, a lack of terminal growth, and wilted, yellow leaves. The rotted root system cannot adequately supply the aboveground portion of the tree with water and minerals. Eventually leaves drop, terminal shoots die back, followed by death of the tree. Often a grower mistakes these symptoms for water stress and irrigates the orchard more frequently; this, in turn, favors further disease development and increased incidence of infected trees. The bark and outer wood of infected roots are reddish brown. The discoloration extends into the wood, and an active canker has a zonate margin. Infected roots are firm and brittle. Aboveground symptoms may be confused with those caused by oak root fungus, but *Phytophthora* does not produce mycelial plaques under the bark, rhizomorphs on the surface of affected roots and crowns, or mushrooms at the base of infected trees.

Trees with *Phytophthora* crown rot infections usually decline rapidly. Trees invaded in fall may be completely girdled by crown rot and killed before new growth starts in spring. These trees have enough stored energy to blossom and leaf out, but when the weather warms up and water is required from the roots, they quickly die. The bark at and below ground level is brown, dead, and often gum soaked; underlying wood is discolored.

Poria wood rot sometimes occurs as a secondary disease in trees with crown rot. White fungus growing on the crown area of the tree indicates the presence of *Poria* (see page 99).

Phytophthora trunk cankers are irregularly shaped; their margins are encircled with concentric lines. They are frequently confused with Ceratocystis cankers. *Ceratocystis* can enter the tree only through bruise-type wounds on the trunk and scaffold branches. Association of bark damage with Phytophthora cankers is purely coincidental. Although *Phytophthora* will readily invade pruning wounds in the fall, winter, or early spring during periods of heavy rainfall, it does not need a wound to enter a tree. Phytophthora trunk cankers grow rapidly,

JOHN MIRCETICH

Phytophthora crown and root rot can infect large numbers of trees in an orchard. It may be difficult to discern any pattern in the distribution of the disease; infected trees may occur singly or in groups.

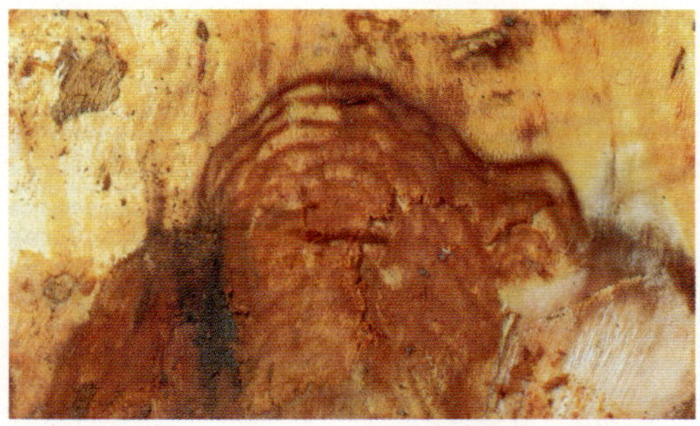

Margins of Phytophthora cankers show a zonate pattern.

Discolored wood in the crown area of this tree indicates a Phytophthora crown rot infection.

***Phytophthora* fungi has directly penetrated the trunk of this tree, resulting in a trunk canker. Suckers at the base of the tree indicate that the roots have not been infected and remain alive.**

Phytophthora branch cankers may occur anywhere along a scaffold.

Frequently, Phytophthora branch cankers occur at pruning wounds.

often 2 to 4 feet per year, while the growth of a Ceratocystis canker is slow. Also, Ceratocystis cankers do not extend below the soil line, whereas Phytophthora cankers frequently extend into the crown and main lateral roots. Both *Phytophthora* and *Ceratocystis* can be present in the same tree, however.

Branch cankers, frequently called aerial Phytophthora, become evident when numerous gum balls appear in groups on a scaffold or at pruning wounds. Gumming occurs on the outer surface of the cankered area. Under the bark, the margin of an active canker consist of reddish concentric rings.

Occasionally, nursery trees are girdled by a canker on the trunk. This disease is known as "bundle rot" and is caused by a species of Phytophthora. When almond trees are dug up, tied together in bundles, and heeled-in before planting, *Phytophthora* infections can occur and spread to a large number of trees if conditions are wet and heeling beds are infested with *Phytophthora*. Often these infections may not become apparent until after the tree is planted in the orchard. Symptoms, including gumming along the trunk, are sometimes confused with bacterial canker infections, but bacterial canker is not known to occur in first leaf trees.

Seasonal Development. All *Phytophthora* species that attack almond trees are soil-borne fungi with a wide geographical distribution, although not every orchard is infested nor do all species occur in all soils. Fungi can persist in soils without a host for a long time but require free water to reproduce and infect the host plant en masse. Thus, crown and root rots frequently occur in orchards with fine-textured soils that are poorly drained or have low areas where water collects. Other sources of inoculum for *Phytophthora* infections include decaying, infected roots and active cankers. Epidemics of the disease occur in years when rainfall is plentiful in late fall, winter, and early spring, and the soil remains flooded or saturated with water for extended periods.

Phytophthora fungi do not require wounds to infect a tree. The pathogen most often enters the tree at sites where infested, waterlogged soil or flood water touches the root, crown, or trunk. Once in the root or crown, the infection, depending on the species of *Phytophthora* involved, may extend into the crown, trunk, or branches. If *Phytophthora* spores are present in water that is used to sprinkler-irrigate the orchard, the pathogen may also enter the trunk or branches at sites where water collects.

The cool, moist conditions that favor Phytophthora root rot generally do not last more than a few months. When conditions are less favorable for the fungi, trees infected with *Phytophthora* species that cause primary root rot often are able to regenerate roots and recover from the damage. Recovery depends on good soil water management and rootstock resistance to the disease.

Infections in both the crown and trunk can damage trees substantially in a short time if conditions are favorable. Crown infections, however, are usually more destructive than trunk cankers because trunks rapidly warm up in spring to temperatures that no longer favor disease development. Most Phytophthora trunk cankers are chronic, but the fungus does not grow during periods of hot or cold temperatures.

One species of *Phytophthora* that causes branch and trunk cankers, *Phytophthora syringae*, is active only in cool weather (50° to 70°F). It may be spread by windblown rain and enters primarily through pruning wounds but may possibly penetrate the bark directly. With the onset of hot weather, the fungus dies within the canker and the affected area can heal over.

Management Guidelines. Proper water management is the most important aspect in controlling root and crown rot. Water should not be allowed to accumulate or stand around crowns of trees. Low spots in the orchard, areas that flood frequently, and places where water penetration is extremely poor should be provided with adequate drainage or left unplanted. Once you have *Phytophthora* in your orchard, chances are the pathogen will remain: eradication is impossible. Be careful not to introduce the pathogen into uncontaminated soil by infected plant material, infested irrigation water, or infested soil on farm equipment.

If you are replanting an area where *Phytophthora* is present, plant trees on small mounds, as shallowly as possible, or on broad ridges with the upper roots near the soil level. Establish berms before planting; the ridges should be 8 to 10 inches high (20 to 25 cm). Planting depth after settling should be no deeper than in the nursery, and the graft union should be well above the soil line.

Use a rootstock that has some resistance to *Phytophthora.* Of the rootstocks used in almond orchards, almond is the most susceptible to *Phytophthora* crown and root rot, followed by almond X peach hybrid, Nemaguard, and Lovell. Marianna 2624 is the most resistant of the rootstocks used for almonds.

To protect the trunk and scaffolds from infection, use low angle sprinklers and remove weeds from around the base of the tree. Weeds tend to maintain high humidity and provide a favorable environment for disease development.

No chemical controls are currently available for this disease, although some are being evaluated for control of Phytophthora root and crown rot in nonbearing almond trees. For further information, consult your farm advisor.

Crown Gall

Agrobacterium tumefaciens

The bacterium that causes crown gall is widespread and occurs in most agricultural soils throughout the world. Although *A. tumefaciens* can cause galls in established orchard trees, it is most damaging to young trees, especially nursery stock.

Symptoms and Damage. Although almond trees do develop galls visible at ground level, most galls develop on the almond root system and cannot be detected unless the soil is removed. Aboveground symptoms of trees infected with crown gall can include open canopies, lack of vigor, yield reduction, lack of terminal growth and small leaves. Although young trees are usually not killed by this disease, they never become very productive. Older trees can usually tolerate crown galls.

Galls are commonly the size of a baseball, but can vary from microscopic to more than 12 inches in diameter. Wood inside a gall lacks growth rings and is softer than healthy wood, but the color is the same as healthy wood. The outside color of the gall is the same as normal bark. Young galls have a smooth surface that becomes rough and irregular as they enlarge and slough off outer layers. Portions may eventually decay and provide entry points for wood-rotting fungi.

Severity of damage caused by crown gall depends upon several factors, including the number and location of infections, size of the galls, age of the tree, and establishment of secondary infections. Infections that develop during the first 3 years after planting are much more debilitating than those beginning later in the tree's life. Older trees can usually tolerate these galls if they are not on primary roots.

Seasonal Development. When galls on infected trees are broken, sloughed off, or moistened, crown gall bacteria can enter the soil, where they will survive for a short period. *Agrobacterium tumefaciens* can be spread with contaminated soil on equipment or with infected plant material. The pathogen infects trees only through wounds, such as those caused by cultivation, pruning, frost injury or growth cracks.

Nursery trees are particularly prone to infection because of the many potential injuries during handling. The pathogen does not travel far inside the tree, and each gall is believed to be a separate infection site.

Management Guidelines. Treat your trees for crown gall just before planting them in the orchard. Dip roots for 30 seconds in a protective solution of *Agrobacterium radiobacter,* a commercially available biological agent (Galltrol). Galltrol provides a protective shield against the crown gall bacterium; it does not eradicate existing galls. A few strains of *A. tumefaciens* are occasionally encountered that are more difficult to control with Galltrol. Research is currently underway to develop better control measures for these strains.

Trees that are heeled-in and root pruned may require dipping before each operation. If only one treatment with Galltrol is economically feasible, use it before planting the trees in the orchard.

The following measures will also help to prevent crown galls: 1) purchase young trees from a nursery that follows recommended procedures for gall control; 2) plant trees with a minimum of handling and avoid injury to trees at planting and during cultural practices involving machinery; 3) remove soil from around the roots before

Crown and root galls are most damaging to the rootstock of young almond trees. Galls disrupt root functions and severely debilitate the trees.

planting, and cut off damaged or torn roots before dipping the root system as described above.

Most of the almond rootstocks are susceptible to crown gall bacterium, but Marianna 2624 is more resistant to invasion by secondary wood-rotting fungi than the other rootstocks.

If a tree is infected with crown gall, you can paint the gall tissue while the trees are dormant with a bactericide registered for this purpose. Expose the gall by carefully removing soil with a stream of water or compressed air. If you use shovels to remove the soil, use considerable care to avoid further injury to the tree. Do not treat more than 50% of the tree's circumference with the bactericide at one time.

Oak Root Fungus
Armillaria mellea

Oak root fungus, also known as Armillaria root rot and shoestring fungus, occurs in a wide range of woody and herbaceous plants throughout temperate and tropical climates. In California the disease is more prevalent in the Sacramento Valley and in areas where oak trees once grew. The fungus can survive for many years in old, infected pieces of wood or root below ground, particularly along former stream beds and near creeks where the soil is moist. The disease is generally limited to areas that are favorable to growth of the fungus, and generally spreads slowly within an orchard.

Symptoms and Damage. Symptoms may not develop until after the disease is well established. The first symptoms of oak root fungus are poor growth or dieback of shoots; small, yellow leaves; and premature leaf drop. The pathogen is able to invade the roots and crown when its dark, shoestringlike structures, called rhizomorphs, come into contact with healthy, underground tissue. Gradually it girdles the crown region and destroys the entire root system.

Rhizomorphs grow on the underground surface of the bark and eventually penetrate living tissue to form fan-shaped, creamy white plaques of mycelium between the bark and woody tissue. These plaques have a strong mushroom odor, and the wood beneath is firm but wet looking. The exterior of the bark usually appears normal, although there may be slight gumming. *Armillaria* may form clusters of mushrooms at the base of infected trees a few days after heavy rain in the fall. The mushroom caps range from 1 to 5 inches (2 to 12 cm) in diameter, are honey yellow and sometimes covered with brown scales; the gills underneath are light brown.

Poria wood rot is often confused with oak root fungus. *Poria* invades only through dead tissues, such as old crown galls, injuries, or dead roots, and the wood it attacks becomes soft and crumbly. White fungal growth in the crown area of the tree on the outside of the bark represents the fruiting structures of *Poria. Poria* does not produce rhizomorphs, fan-shaped plaques between the bark and wood, or mushrooms.

Occasionally oak root fungus may cause the tree to suddenly collapse during the dry summer months—an occurrence that may be confused with Phytophthora crown rot. The presence of rhizomorphs, fan-shaped plaques, or mushrooms will confirm oak root fungus as the cause of death.

Seasonal Development. *Armillaria* develops most rapidly in moist, cool soil. The fungus spreads not by spores, but by the growth of rhizomorphs. When tree roots from a healthy tree come in contact with diseased roots, rhizomorphs can penetrate the healthy roots. No wounds are necessary for infection. Clusters of trees may be infected at one or several sites in the orchard.

The fungus can survive for many years in dead or living roots of fruit and nut trees, oak, aspen and several fir and willow species. The disease often occurs along former stream beds, near creeks where soil is moist, and where dead roots or stumps harbor the fungus, especially in orchards previously infected with the disease. Infection and disease development occur mostly during spring and early summer. Trees usually die within 1 to 3 years, depending upon the susceptibility of the rootstock. Death is hastened by summer heat and inability of the tree to obtain sufficient water from the soil.

Management Guidelines. Management of oak root fungus relies primarily on preventing infection of new trees. Once infection occurs, very little can be done to save a tree. You can reduce chances of infection by carefully preparing planting sites for new orchards, by practicing good sanitation, and by detecting diseased trees early.

Avoid planting in a site infested with *Armillaria.* When this is not possible, remove and burn at the infestation site all roots larger than 1/2 to 1 inch (1.2 to 2.5 cm). Smaller infected roots will decay rapidly, and the fungus cannot survive without a host in the soil. If infected roots are moved elsewhere for burning, some may drop and establish new infection sites.

Let the soil dry out by withholding water during the summer for one to two seasons to prepare it for fumigation; this will improve penetration of the fumigant and also kill some of the fungus. Plant deep-rooted, nonirrigated cover crops such as mustard, sudangrass, or safflower to help reduce soil moisure. Deep-till the area after drying. Warm, dry, well-worked soil will greatly improve the ability of the fumigant to penetrate the soil. The deeper the fumigant penetrates, the more effective the treatment will be. The best time of year to fumigate for oak root fungus is in September before the rains begin. This recommendation is specific for oak root fungus control. To fumigate for nematodes and other organisms that inhabit the surface layers, moist soil is essential.

After fumigation, aerate the soil 1 to 2 months before planting. Be aware that fumigants do not always kill the fungus in all infected roots, especially those deep in the soil. Retreatment may be necessary in localized areas. Consult your farm advisor for advice on the type of fumigant to use.

In areas where oak root fungus is a major problem, choose a less suceptible rootstock. Marianna 2624 rootstock offers some resistance, but some cultivars, including Nonpareil, are not directly compatible with it. Jordanolo, Ne Plus Ultra, Peerless, Mission, Merced, Carmel, Padre, Price, Harvey, and Thompson may be grown on this rootstock. Consult your nurseryman for a current list of compatible and incompatible cultivars.

Be careful not to introduce the pathogen into an established orchard. If you have an area infected with *Armillaria,* do not move infected roots from this area to other locations when performing cultural operations.

Check the orchard for symptoms of oak root fungus in late summer when dead or declining trees are most obvious. Look for rhizomorphs on the surface of the bark by carefully removing soil from the crown or large roots. Remove the outer bark with a hatchet or strong knife to locate mycelial plaques.

Once symptoms appear on one tree, remove infected trees and adjacent, apparently healthy trees; the disease is likely to have already spread to the roots of surrounding trees.

If you remove the bark of a tree infected with *Armillaria,* white mycelial plaques will be evident at or just below ground level.

***Armillaria* plaques grow on the surface of the wood in a fan-shaped pattern.**

Brittle, black rhizomorphs of *Armillaria* can be difficult to find because they tend to blend in with the surface of the root or crown on which they are growing. They are easily broken if care is not taken when removing soil from these areas.

Wood Rots

Of the several species of wood-rotting fungi that can affect almond trees, the most common is Poria wood rot. Poria wood rot, caused by *Poria ambigua,* also is found on cherry, peach, and apple. The fungi reside on decaying

White fungal growth at the base of a tree is usually the first indication of Poria wood rot.

wood and usually invade root and lower trunk areas previously infected with crown gall or crown rot.

Poria wood rot appears first as a white fungal growth around the base of an affected tree in late summer. The growth, which is made up of *Poria* fruiting structures, extends for a short distance up the trunk and out into the soil. By the time the white fruiting structure appears, the wood-rotting fungus is well established in the inner structural tissues of the tree. Although the tree is still productive and may appear perfectly healthy otherwise, it will probably fall over during a windstorm because the interior wood of the tree is weak, soft, and decayed.

Other wood-rotting fungi may also affect almond trees, including one that is common in older orchards and thought to be a species of *Heterobasidium (Fomes)*. This wood rot is similar to *Poria* in that it rots the inner structural wood, and the tree falls over in a windstorm. Instead of producing a white fungal fruiting structure, however, *Heterobasidium* produces shelflike, protruding conks on the tree trunk.

Most wood rots are secondary diseases that only invade injured or dead tree tissue. The best way to protect a tree from wood rot fungi is to reduce crown and root problems with careful soil and water management and to avoid cultivation injuries. No chemical treatments are recommended for wood-rotting fungi, and destroying the conk or the white fungal growth at the base of the tree is useless—the fruiting bodies are only an indication of extensive inner rot.

When a tree falls over and is removed, no treatment is necessary for *Poria* or other wood rots before planting another tree in the same spot. If other disease organisms or nematodes are present in the soil, however, preplant fumigation may be necessary.

TRUNK AND BRANCH CANKERS

A canker is an area of necrotic or dead bark tissue. It can be any size. Most cankers are found on large branches, scaffolds or trunks, but they also can occur on small, young branches. On the outside of the tree, a canker usually appears as a depressed area of bark that is somewhat darker in color than the surrounding healthy bark tissue. All cankers have definite margins where healthy and necrotic tissue meet. Usually gum balls are produced at the margin or over the surface of cankered areas.

Cankers may be caused by physical injuries such as sunburn, mechanical injuries such as shaker or mallet wounds, or by plant pathogens such as bacteria and fungi. Cankers described in this section are those caused by plant pathogens. Distinguishing characteristics of each canker are outlined in Table 12 and described in the following sections. In addition, several species of *Phytophthora* that cause trunk and branch cankers are described above in the section on Phytophthora crown and root rot.

Ceratocystis Canker
Ceratocystis fimbriata

Ceratocystis or mallet wound canker is a widespread disease in almonds. Cankers develop in areas of the trunk or branches that have been damaged by cultural equipment or by rubber-covered mallets or mechanical shakers during harvesting.

Cultivars most susceptible to the disease are Mission, Ne Plus Ultra, and Nonpareil.

Symptoms and Damage. Ceratocystis cankers first appear as elongated water-soaked areas with amber-colored gum at the canker margins. Infected tissue turns dark brown, and eventually the affected area becomes sunken. Dark stains may permeate the heartwood and extend longitudinally 20 inches (50 cm) or more past the margin of the canker on the bark. Cankers can girdle and kill infected limbs or trunks. A limb 4 to 6 inches in diameter can be girdled in 3 to 4 years. However, some cankers become inactive and heal after two or three growing seasons. Trees consistently damaged by harvesters year after year are the ones most severely affected by this canker.

Cankers caused by *Ceratocystis* fungus may be confused with branch cankers caused by *Phytophthora* spp. Gumming occurs with both types of cankers but usually in a different pattern. Gum balls generally occur at the margins of a Ceratocystis canker, whereas in Phytophthora cankers, they generally occur throughout the diseased area. Unlike Phytophthora cankers, which can develop on uninjured portions of the tree, cankers caused by *Ceratocystis* are always associated with mechanical injuries to the bark. Also, Ceratocystis cankers grow slowly from year to year, whereas those caused by *Phytophthora* advance rapidly. (Photos on page 102.)

Seasonal Development. *Ceratocystis* fungi are spread by several species of sap-feeding beetles and a fruit fly. These insects feed on the fungi in diseased trees, including other stone fruit trees, and either ingest fungal spores and later excrete them, or come into contact with the spores and transport them on their bodies to new locations. Insects that have fed on spores as larvae can retain the fungus through pupation and emerge with it as adults. Adult insects can also be contaminated by the fungus in winter months if they inhabit old bark wounds.

Following a bark injury, most almond trees are susceptible to *Ceratocystis* infections for 8 to 14 days. If a fungus-carrying insect is attracted to the wound, the fungus infects bark and exposed cambium tissue. Once it is on exposed cambium, the fungus can invade healthy bark tissue and young xylem tissues. Dark stains permeate the sapwood and heartwood, but the fungus seldom penetrates further than the xylem of the previous year's growth.

Management Guidelines. The most effective way to prevent Ceratocystis canker is to avoid shaker injury to trunks and scaffolds. If the bark is injured, shave the rough portions to promote callus formation. Do not irrigate orchards near harvesttime. Shakers are more likely to damage recently irrigated trees because the bark slips more easily.

To control established cankers, you can perform tree surgery. Cankers are not easy to remove, however, and if you miss some of the infected tissue, the fungus can continue to grow and the canker will return. More often than not, tree surgery must be done year after year until you finally remove all infected tissue, and it may not be economically feasible. The best time to carry out tree surgery is from December to February; both the *Ceratocystis* fungus and the insect vectors are less active at this time. (However, surgery at this time of year may provide an opening for infection by aerial *Phytophthora* fungi.) Remove infected bark and 1/4 to 1/2 inch (6 to 13 mm) of the woody tissue underneath the bark. Extend the cut at least 1 inch (2.5 cm) beyond the visible canker margin. Recheck the area the following year. If the canker has returned, repeat the process.

If surgery is done in the winter months when few insects are active, leave the wound undressed. The value of putting dressings on tree wounds following surgery has not been established, and it slows the healing process.

Avoid shaking areas of the tree that have had surgery as this can reopen wounds.

Bacterial Canker
Pseudomonas syringae

Bacterial canker, also known as gummosis or sour sap, occurs on all commercially grown stone fruits. In almonds, the disease is more common in areas with coarse-textured or sandy soils than in areas with fine-textured or heavy clay soils.

The bacterium, *Pseudomonas syringae,* is commonly present on the surfaces of many plants and causes both bacterial canker and bacterial blast (see page 109) in almonds. No consistent correlation exists between the presence of bacterial canker and blast; orchards with one of the diseases may or may not have the other. Evidence suggests, however, that freezing temperatures, in addition to favoring blast, may also enhance the development of cankers.

Symptoms and Damage. In almonds, bacterial canker can either cause a limited canker or a more severe form of the disease known as "sour sap." Sour sap generally occurs in younger trees, and limited cankers occur in older

Table 12. General Characteristics of Almond Cankers.

	Bacterial canker *Pseudomonas syringae*	Ceratocystis canker *Ceratocystis fimbriata*	Dothiorella (band) canker *Botryosphaeria dothidea*	Foamy canker *Zymomonas* sp.	Phytophthora branch canker (aerial Phytophthora) *Phytophthora syringae*	Phytophthora trunk canker *Phytophthora* spp.
Casual agent	Bacterium	Fungus	Fungus	Bacterium	Fungus	Fungus
Period of activity	Winter	Summer	Summer	Summer	Fall Winter Spring	Summer
Annual or chronic	Annual	Usually chronic	Not strictly annual	Annual	Annual	Usually chronic
Type of margin	Flecking	Zonate	Zonate	Not established	Zonate	Zonate
Gumming	Occasional gum balls in cankered area	Gum balls around edge of canker	Gum balls in band arcund limb or trunk	Oozing reddish gum and white foam	Clustered gum balls, usually around a pruning wound	Numerous random gum balls throughout cankered area

A shaker wound at the base of the scaffold on the left side of this tree provided an entry point for *Ceratocystis*. The resulting canker has girdled and killed the entire scaffold.

This young tree was killed during the winter by the sour sap phase of bacterial canker. Because only the aboveground portion of the tree is affected, roots put out suckers in the spring.

Although this bacterial canker has crossed the graft union, it is restricted to the aboveground portion of the tree. The reddish brown color of the diseased wood is characteristic of a *Pseudomonas syringae* infection.

R.M. BOSTOCK

Ceratocystis cankers are always associated with bark injuries. Generally, gumming occurs at the margins of the canker.

The dead spur on this branch served as the entry point for the developing bacterial canker, shown by the gumming just to the left of the spur.

Removing bark from an infected branch reveals the extent of a bacterial canker infection.

Active bacterial canker infections have flecks of reddish brown wood just beyond the margin of the main canker. The flecks signify that the canker is expanding and also aid in distinguishing bacterial canker from most fungal cankers.

trees or trees of any age that are infected in late dormancy. In epidemic years heavy tree losses have occurred, especially in orchards where the trees are 2 to 7 years old. Older trees are not usually seriously damaged, and trees less than 2 years old are rarely attacked by this pathogen. First leaf trees showing similar girdling and gumming symptoms may have bundle rot, a disease caused by *Phytophthora* spp. (see page 96).

Bacterial cankers can occur on trunks, scaffolds or small branches. Inactive, healed cankers are rough,

Foamy canker produces copious amounts of reddish gum that drain down the scaffolds and trunk of the tree to form a puddle on the ground.

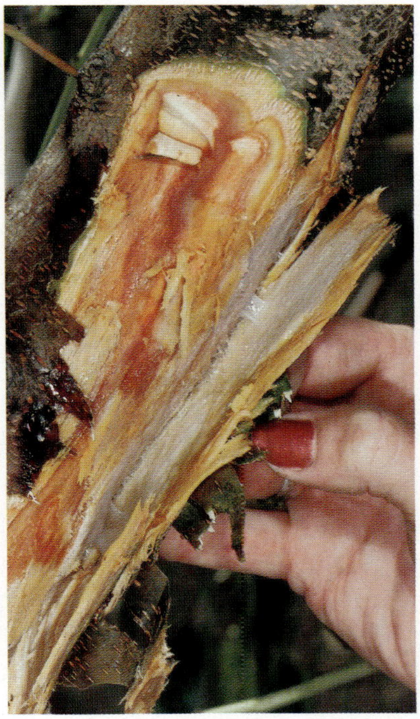

Foamy canker foam is produced under the bark in the rotted and mushy cambium layer.

Band canker produces a band of gum balls around the trunk or affected limb.

Leaves remain throughout the growing season on limbs killed by foamy canker. From a distance the disease may resemble Verticillium wilt.

Underneath the bark, band cankers grow around the trunk instead of longitudinally as do cankers caused by other pathogens.

elongated, and sunken into the bark. Active bacterial cankers usually are not visible externally unless there is some gumming. Internally, the bark and outer layers of wood are dark, reddish brown in the canker area. Often, small brown flecks in the inner bark beyond the canker margin occur in active cankers. These dark spots represent new infections caused by the bacteria as they move out to infect healthy wood. The irregular shape of the bacterial canker distinguishes it from most fungal cankers, which usually are symmetrical in shape and have contoured margins. Although bacterial cankers are active during dormancy, they are not apparent until early spring when they gum or produce amber-colored sap; the name "gummosis" refers to this symptom. Branches that are girdled by a canker may die and "sour out," producing a brownish watery ooze on the surface of the bark.

Sour sap occurs when a tree is invaded by *Pseudomonas syringae* early enough in dormancy to allow the bacteria sufficient time to kill a large branch or the entire aboveground portion of the tree. Generally, the younger and smaller a tree is, the more likely it is to be completely invaded by the bacteria. The dead wood is brown, moist, and sour smelling. In the spring, when other trees bloom and leaf out, diseased trees produce a watery sap that runs down the tree and has a vinegarlike odor. Invasion by the pathogen can cross the graft union, but it is restricted to aboveground portions of the tree. The tree may produce suckers from the base of the trunk, but the entire aboveground portion of the tree is dead.

Seasonal Development. Bacteria survive the summer on leaves and bark of trees. In fall and winter the pathogen is moved by rain and enters dormant trees through buds or natural openings such as leaf scars. The disease progresses throughout the dormant season forming cankers or causing souring out. When the tree resumes active growth in spring, it forms callus tissue around the canker and the activity of the bacterial canker pathogen is suppressed. In rare instances, the canker may not be completely encircled by callus tissue. In this case, the bacterium may become active the following winter and the canker continue to grow. Also, bacteria may occasionally remain alive throughout the summer in small areas outside the margins of a canker and resume growth in the fall.

Many factors appear to influence the tree's susceptibility to the canker phase of the disease. In general, bacterial canker is less damaging on vigorous trees. Freezing temperatures appear to enhance development of bacterial cankers, and trees between 2 to 7 years old have a higher incidence of these cankers than do older trees. Trees are more susceptible if grown on coarse-textured, shallow, or nitrogen-deficient soils, or on soils containing a high population of ring nematodes, *Macroposthonia xenoplax.* Late fall irrigation increases the severity of bacterial canker in young peach trees grown in coarse-textured soils, but no research has been done to determine if the same is true for almonds.

Trees infected early in dormancy are more seriously damaged than trees infected as spring approaches. The earlier the invasion occurs during the dormant season, the more time the bacteria have to extend the infection before active tree growth resumes. If active growth of the tree in the spring is impaired by other diseases or environmental stresses, the development of bacterial canker will be enhanced.

No evidence is available on whether the time of pruning plays a role in the disease cycle, or if insects are involved in either infection or dissemination of the pathogen. Rain appears to play a major role in the dispersal of the bacteria to natural openings in the tree.

Management Guidelines. Research on other stone fruits indicates that preplant fumigation can reduce the incidence and severity of bacterial canker, especially in coarse-textured soils and shallow soils above a hard pan. Preplant soil fumigation controls ring nematodes, which may predispose trees to bacterial canker. Newly planted trees can be protected from ring nematodes by preplant fumigation for 2 to 3 years—the period when they usually are most susceptible to the disease.

More information is needed on rootstock selection for almonds grown in areas where bacterial canker is a problem. Research on prune trees indicates that trees grown on Lovell and Nemaguard peach roots suffer less damage from bacterial canker than do those grown on Marianna 2624 rootstock.

Generally, vigorous trees have less trouble with bacterial canker. Practice measures, such as adequate fertilization and irrigation, that encourage good plant growth.

Prune severely cankered or dead scaffold limbs to discourage secondary invasion by wood rot organisms. Experiments on the effectiveness of copper sprays have yielded no data that suggest this approach works.

Foamy Canker
Zymomonas sp.

Foamy canker, a relatively new disease of almond, was first reported in 1974. Although it occurs throughout the San Joaquin Valley and more recently in Butte County, foamy canker generally affects only a few scattered trees within an orchard. The disease has been found primarily on second- and third-leaf trees of the Carmel cultivar, but it also occurs on healthy or weakened trees of various cultivars and ages. No particular pattern or cultural practice has been associated with the disease, nor has this disease been reported in areas outside the Central Valley.

Symptoms and Damage. When the canker is active, copious amounts of watery, reddish gum drain down the scaffolds and trunk to form a puddle on the ground. In addition, white foam cascades down the tree on the surface of the gum. The foam bubbles forth from the bark and resembles the foam of beer. It has an alcoholic odor that suggests fermentation is taking place. Under the bark the cambium, or outermost layer of wood, is rotted, white and mushy. Ultimately, bark and wood die and turn dark brown. (Photos on page 103.)

Like bacterial canker, foamy canker produces very irregularly shaped lesions. Diseased tissue may completely surround a healthy area leaving islands of tissue that are still capable of producing shoots. These shoots don't live very long however. The disease is active only in the growing season, and cankers heal over in the winter. If the tree trunk or limbs are girdled by the disease, leaves die and remain on the tree. From a distance, this symptom resembles Verticillium wilt, but unlike Verticillium, the lower trunk and rootstock remain unaffected and produce vigorously growing suckers. Foamy canker might be confused with bacterial canker, but bacterial canker is active in the dormant season, not in the summer.

Seasonal Development. Foamy canker usually begins at the scaffold crotch of the tree and advances up into primary and secondary limbs and down toward the bud union. It does not progress past the bud union, and the roots are not affected. Symptoms first appear in late July. Disease activity usually tapers off in fall, and the canker becomes inactive. Most cankers do not reactivate the following spring. Warmer than average temperatures may be connected to the occurrence of this disease.

Management Guidelines. Young trees that aren't killed by the canker can be pruned and retrained, or removed. If only a portion of an older tree has collapsed, prune out the dead wood. Make your cut below the dead tissue, in healthy wood. No chemical controls are available.

Band Canker
Botryosphaeria dothidea

Band or dothiorella canker occurs sporadically throughout the Central Valley. It primarily attacks trees in their third to fifth year. It has most commonly been reported on vigorous Nonpareil trees and also on Mission, Ne Plus Ultra, Davey, Drake, and Carmel. (Photos on page 103.)

This canker is unusual because unlike other cankers it extends around the branch or trunk instead of longitudinally along the affected part. Cankers usually occur on the upper part of the trunk or the lower portion of scaffold branches, with infection taking place through growth cracks in the bark. Fungus kills the bark and cambium layer, and the affected area becomes sunken and frequently girdles the limb. During the growing season copious amounts of amber-colored gum exudes from the cankered area forming a necklace or band of gumballs around the affected part of the tree. Band cankers occur infrequently, are active only during warm summer weather, and usually do not reactivate the following year. For these reasons, band canker is of minor importance in almonds.

VASCULAR SYSTEM DISEASES

Pathogens that attack the vascular system of a tree plug the xylem vessels and halt the movement of water and nutrients up from the roots. As a result, leaves yellow and die on one or more branches. Usually the dead leaves remain on the tree throughout the growing season.

Almond Leaf Scorch,
An Unidentified Bacterium

Almond leaf scorch, or golden death, was first discovered in Mendocino County in the 1940s; soon after, the disease was found in the Lancaster area of Los Angeles County and the Brentwood-Antioch area in Contra Costa County. Almond leaf scorch is also found in a few scattered locations, mostly as single affected trees, in the San Joaquin and Sacramento valleys. Although the disease is not likely to become a threat to almond production in these areas, there is always a possibility of a major flare-up. Almond leaf scorch remains a serious threat to almonds grown in the Contra Costa area.

The bacterium that is responsible for almond leaf scorch also causes Pierce's disease in grapes and alfalfa dwarf. It is spread from plant to plant by spittlebugs and sharpshooter leafhoppers that feed in the xylem.

Symptoms and Damage. Leaves on infected trees appear normal in spring, but develop symptoms in early to mid-July. Tips and margins of leaves turn yellow, then brown, and a golden yellow band forms between the green and brown leaf areas; the intensity of this yellow band varies with cultivar. Affected leaves do not wilt, but eventually the entire leaf blade turns brown and necrotic. Some leaves may drop prematurely, but many remain on the tree until the end of the growing season.

Newly infected trees typically have one terminal branch that is affected. But if you examine the trees closely, several developing infections on the same or nearby scaffolds will be apparent. As the disease progresses from year to year, trees have reduced terminal growth, large numbers of dead spurs and small branches, and a dieback of terminal shoots. Within 3 to 8 years, the disease spreads over the entire tree, yields are greatly reduced on affected branches, and eventually the tree dies. No obvious effects of the disease can be seen in the wood without a microscope, but if you soak cross sections or slivers of longitudinal sections of diseased wood from young branches in acidified methanol for 30 to 60 minutes, dark red streaks and splotches will develop. To prepare this solution, add 1 teaspoon (5 ml) of concentrated hydrochloric acid to 16 ounces (500 ml) of 100% methanol.

Seasonal Development. Almond leaf scorch is caused by a bacterium that invades xylem tissues of almond trees and produces a suspected toxin that causes leaf burn. The bacterium is believed to reside in a wide array of host plants, most of which do not exhibit symptoms but may act as reservoirs for the bacterium.

Spittlebugs and several species of leafhoppers, called sharpshooters, transmit the pathogen that causes almond leaf scorch. These insects become carriers of the bacte-

Foliage on trees infected with almond leaf scorch will gradually turn golden in mid-summer. Leaves may drop prematurely, but many remain on the tree until the end of the growing season.

The lower leaf in this photo is just beginning to show symptoms of an almond leaf scorch infection. In the upper leaf, the zonate pattern typical of this disease is just beginning to appear.

As leaves on trees with almond leaf scorch become necrotic, the zonate pattern in the scorched area becomes more prominent.

rium when they insert their mouthparts to feed in the xylem of an infected plant. An almond tree is infected when an insect that carries the bacterium feeds on it. Occurrence of almond leaf scorch depends upon amount and type of vegetation near the orchard, susceptibility of the cultivar, and environmental conditions that are favorable to high populations and dispersal of sharpshooters and spittlebugs.

Management Guidelines. In areas where almond leaf scorch is prevalent, annually inspect orchards in mid- to late July. Prune out all infected limbs in newly infected trees, 2 to 3 feet below symptoms. Do this as soon as possibly-infected shoots are detected. In heavily infected areas, replant with cultivars such as Harvey, Ruby, Padre, Solano, Carmel, Butte, and Fritz, which have shown resistance in field and laboratory tests. Susceptible cultivars include Nonpareil, Ne Plus Ultra, Jordanolo, Milow, Peerless, Long IXL, and Trembath.

The most common sharpshooter vector of almond leaf scorch is the green sharpshooter.

Color patterns of spittlebugs, another vector of almond leaf scorch, are highly variable, but all spittlebugs can be recognized by their general form and shape.

Verticillium Wilt
Verticillium dahliae

Verticillium wilt, or blackheart, is caused by a soil-inhabiting fungus that is widespread in soils throughout temperate regions of the world. It attacks a wide variety of fruit trees and field crops such as cotton, tomatoes, and potatoes. Verticillium wilt often becomes a serious problem in cultivated soils where the presence of suitable host plants allows the fungus to multiply. Young trees, especially those of the Ne Plus Ultra and Carmel cultivars, are highly susceptible to Verticillium wilt. All of the rootstocks used in almonds are highly susceptible except Marianna 2624. Cool spring weather and high soil moisture favor disease development.

A Verticillium wilt infection usually becomes apparent when leaves on one or more branches suddenly wilt, turn light tan, and die. Dead leaves generally remain on the tree throughout the growing season.

Symptoms and Damage. The first evidence of Verticillium wilt in almonds is sudden wilting of foliage on one or more branches in late spring or early summer. Leaves turn light tan and die, although they often remain on the tree. Sometimes the lowest leaves on a branch will wither and fall. The pathogen will gradually kill other leaves as it moves up into the branch. Internally, the xylem of the wood turns dark, and cross sections of the branch will show a full or partial ring of discolored tissue. This darkened tissue extends down the affected scaffold through the trunk and to its point of origin in the root system. Symptoms typically affect one side of the tree, depending upon which portion of the root system is infected. Several scaffolds on a tree may be involved, however, if the pathogen infects a large portion of the roots.

A cross section of a branch infected with Verticillium wilt shows a full or partial ring of discolored tissue.

Two- or three-year-old trees are usually more severely affected than older trees. Verticillium wilt causes poor growth and low productivity but does not usually kill almond trees. Terminal growth on infected and previously infected branches is poor, and these branches are most likely to be reinfected by the fungus the following year, resulting in further stunting. Once trees are 4 to 5 years of age, they recover from the disease. By this time, almond tree roots have penetrated below the top 3 feet of soil, where most of the inoculum resides, and trees may have developed some tolerance to the disease.

Discolored vascular tissue due to Verticillium wilt can also be seen by removing bark from an infected branch.

Seasonal Development. The fungus can survive for years in the soil. It enters the tree through young roots, establishing itself in the water-conducting tissues of the plant, and renders the xylem vessels no longer functional. Cool spring weather is most favorable for disease development. High summer temperatures actually kill the fungus in the tree. If new infection through the roots does not occur, trees recover and grow normally the next year. Although the pathogen may reside in roots for a period of time, old infections eventually die. Once roots have penetrated below the level of inoculum in the soil, new

Table 13. Partial List of Crops That Are Susceptible or Resistant to Verticillium Wilt.

SUSCEPTIBLE CROPS		RESISTANT CROPS[1]	
Common name	Scientific name	Common name	Scientific name
cantaloupe	*Cucumis* spp.	asparagus	*Asparagus officinalis*
Casaba melon	*Cucumis* spp.	barley	*Hordeum vulgare*
castor bean	*Ricinus communis*	bean	*Phaseolus* spp.
cotton	*Gossypium* spp.	carrot	*Dacus carota*
Crenshaw melon	*Cucumis* spp.	celery	*Apium graveolens*
eggplant	*Solanum melongena*	corn	*Zea mays*
honeydew melon	*Cucumis* spp.	milo	*Sorghum vulgare* var. *Durra*
mint	*Mentha* spp.	oat	*Avena sativa*
muskmelon	*Cucumis* spp.	pea	*Pisum sativum*
okra	*Hibiscus esculentus*	sorghum	*Sorghum vulgare*
peanut	*Arachis hypogaea*	sweet potato	*Ipomoea batatas*
Persian melon	*Cucumis* spp.	wheat	*Triticum aestivum*
potato	*Solanum tuberosum*		
safflower	*Carthamus tinctorius*		
strawberry[2]	*Fragaria chiloensis* var. *Anarassa*		
tomato[3]	*Lycopersicon esculentum*		
watermelon	*Citrullus vulgaris*		

1. Although other crops, such as alfalfa and lettuce, are resistant to Verticillium wilt, they were not listed because they are good hosts for another almond pest, the root knot nematode.
2. Resistant cultivars of strawberry are Blakemore, Catskill, Sierra, Siletz, Surecrop, Vermilion, and Wiltguard.
3. The VF-numbered varieties of tomatoes are resistant to Verticillium wilt and Fusarium wilt.

infections will not occur and symptoms will virtually disappear from a previously infected young orchard.

Management Guidelines. When establishing a new orchard, try to select a site that has never been used to grow cotton, tomatoes, melons, potatoes, or other highly susceptible crops (Table 13). Intercropping with susceptible crops in young orchards is not recommended.

If the fungus has infected your trees, proper fertilization and irrigation will help young trees to recover from the attack. Prune out all weakened and dead limbs in mid-summer when the fungus is no longer active.

BRANCH, FOLIAGE, AND FRUIT DISEASES

Several pathogens cause infections that are limited to branches, foliage and fruits. They produce localized lesions or larger infections such as shoot strikes. Almond trees are susceptible to infection by many pathogens during bloom because the flowers provide a good entry point. Infections also occur following rains or sprinkler irrigation when water gets into the tree's canopy, spreading the pathogen and creating conditions favorable for infection.

Brown Rot Blossom and Twig Blight
Monilinia laxa

Brown rot occurs in most fruit-growing regions of the world. Most brown rot is caused by *Monilinia laxa*, but some is caused by a closely related fungus, *M. fructicola*. It affects a variety of fruit crops including apricots, almonds, cherries, plums, nectarines, peaches, and prunes. Brown rot first appeared on California almonds in the late 1800s and currently is common in the northern San Joaquin Valley and the Sacramento Valley. Disease development is favored by frequent rains or fogs during bloom. Every almond cultivar is susceptible to brown rot, but Butte, Carmel, Drake, Jordanolo, Ne Plus Ultra, IXL, and Ruby are the most severely affected. Severe blossom infection is less common in Peerless and Davey, and the susceptibility of the Mission cultivar is variable.

Symptoms and Damage. The first symptom of brown rot is the withering of almond blossoms. Withered blossoms remain on the tree and are covered with light brown, powdery masses of fungus spores during humid weather. The fungus moves from the blossoms into adjacent twigs and kills them as well. Gumming may occur at the base of dead flowers and on the bark of twigs at the junction of dead and live tissue. Dead twigs eventually turn a light, yellow brown color. Leaf clusters or shoots that are not adjacent to infected flowers will not be infected, although they may collapse if a blossom that is lower on the branch becomes infected and girdles the twig. In contrast, another disease known as bacterial blast affects both leaf and blossom clusters. Because flowers are destroyed, the net effect of brown rot blossom and twig blight is a reduction in fruit set. (Photos on page 110.)

Infection of green fruit by *Monilinia laxa* can occur but is uncommon in almonds; green fruit is more likely to be infected by two other species of fungi: *Botrytis cinerea* and *Sclerotinia sclerotiorum* (see page 114). *Monilinia laxa* and a closely related fungus, *M. fructicola,* do infect splitting hulls, however, and cause hull rot (page 117).

Seasonal Development. Brown rot fungi survive from one season to the next in infected blossoms, twigs, and fruits. In late winter the fungus begins to develop light gray tufts of spores on the infected surface. These spores are spread by wind or by rain, which splashes them to new infection sites, and they infect blossoms. The critical period for flower infection is from the time unopened flowers emerge from winter buds (pink bud stage) until petals are shed (petal fall). Optimal conditions for infection occur in foggy or rainy weather when temperatures are around 77°F (25°C) or higher. Disease will develop at temperatures as low as 50° to 60°F (10° to 15.8°C), however, if there is sufficient rain.

Management Guidelines. On highly susceptible scion cultivars (see Figure 12, page 21), fungicides are the recommended means of control. In severe cases, a dormant treatment with an eradicant fungicide may be advisable to kill spore pustules and suppress spore development. In most cases, however, fungicides applied at bloom give adequate control. Make your first treatment at the pink bud stage (see back cover) when petals start to emerge beyond the green sepals. Ideally you should apply the fungicide to each cultivar when it reaches the pink bud stage. If this is not possible, time your spray to the pink bud stage of the most seriously affected cultivar in your orchard. In most years, application of a single systemic benzimidazole fungicide, such as Benlate or Topsin M, gives adequate protection. Recent detection of benzimidazole resistant *Monilinia laxa,* however, means these fungicides may no longer be effective in some orchards for controlling brown rot. In cool, wet years the bloom period is often extended, and a second systemic fungicidal spray may be necessary just before full bloom. If you spray with a contact fungicide, you will need to spray at both pink bud and full bloom.

Various cultural practices can be helpful in managing brown rot. If you are planting an orchard in areas where brown rot is prevalent, consider using less susceptible cultivars, and space your trees so that there will be adequate air movement between them. Never sprinkler irrigate trees during bloom, and use low-angle sprinklers in sprinkler-irrigated orchards to prevent water from splashing the spores to new infection sites.

Bacterial Blast

Pseudomonas syringae

Bacterial blast affects all stone fruits but is particularly severe in almonds, apricots, plums, and peaches. Occurrence of bacterial blast is sporadic, depending upon the climatic conditions of a given year. Optimal conditions for this disease are cold or freezing temperatures and rain during the bloom period; the pathogen is spread by water. Cultivars that bloom early, such as Ne Plus Ultra, are more susceptible to bacterial blast because the buds and blossoms are more likely to be damaged by frost or experience frosty conditions.

Bacterial blast is caused by the same pathogen as bacterial canker.

Symptoms and Damage. The bacterium attacks just before, during, or shortly after bloom and kills buds, blossoms, young leaves, and shoots. Blossoms or opening buds turn black and remain attached to the tree. Leaves may develop dark spots that fall out leaving a hole. When the petiole is infected, the leaf is often distorted. Entire spurs and young shoots may be killed, but the bacteria usually do not move very far into older twigs. Dead spurs with dark brown to black leaves and flowers remain in the tree, and these "flags" are an indication of a previous blast infection. Blast can be devastating, and under certain conditions it can cause almost complete crop loss.

Seasonal Development. Wet weather activates the bacterium, which is commonly present on leaves and bark of the tree. In addition, the pathogen has the ability to induce formation of ice crystals at slightly higher temperatures than normal, resulting in increased plant injury. Because it is associated with frost injury, bacterial blast most commonly occurs in low-lying, cold pockets of an orchard and in lower parts of trees.

Freezing temperatures at bud break or during the bloom period injure flower tissue and provide a point of entry for the bacteria. In the absence of freezing temperatures, blast usually is of little consequence. If freezing temperatures occur and the bacteria invade the tree during bud break, bacterial blast occurs in flower and leaf buds, which die before opening. If infection occurs during bloom, blast symptoms will occur on flower blossoms

Brown areas on these petals signal the beginning of a brown rot infection.

BETH TEVIOTDALE

The tip of this shoot was killed by a bacterial blast infection during the time that the tree was leafing out. The dead leaves will remain on the tree throughout the season.

The tip of this twig is being girdled by a brown rot infection. Leaves at the terminal have wilted and will eventually die.

Brown rot fungi can overwinter in twig cankers that were infected the previous year. Reproductive structures of the fungi appear as gray pustules on margins of the canker.

Gumming on this brown rot infected branch marks the junction of dead and live tissue. Note that nearby leaf clusters have not been infected, showing that the problem is brown rot, not bacterial blast.

Brown rot fungi can also overwinter on dead blossoms.

and young leaves. In both cases, twig and spur dieback can also occur. Less commonly, young green shoots are attacked directly and killed. Bacterium ceases to be active in late March, when the tree begins active growth and develops resistance to the pathogen.

Management Guidelines. Because occurrence of bacterial blast is sporadic and unpredictable, chemical controls are not generally recommended. Protecting blossoms from frost may provide control because of the connection between freezing temperatures and incidence of infection. More research is needed before specific management recommendations can be made.

Shot Hole

Stigmina carpophila (Coryneum beijerinckii)

Shot hole is prevalent in the Pacific coast states, as well as other parts of the world, and is a major disease of peaches, nectarines, apricots, and almonds. It used to occur most frequently on almonds growing in the Sacramento and northern San Joaquin valleys, until widespread use of sprinkler irrigation was adopted in the central and southern San Joaquin Valley. The disease is most severe during prolonged, wet, spring weather.

Ne Plus Ultra, Drake, Peerless, Thompson, Jordanolo, and Nonpareil cultivars are more susceptible than Mission.

Shot hole leaf lesions begin as small red freckles. As lesions expand, the center turns brown and is surrounded by a darker margin. Several lesions in the leaf on the lower right are breaking away from the surrounding leaf tissue and will eventually fall out, leaving a shot hole.

The small, dark fruiting structures in the center of these leaf lesions are characteristic of shot hole infection.

Shot hole lesions on fruit begin as small, dark spots that eventually become rough and corky.

Shot hole twig lesions begin as small, purple spots.

Twig lesions serve as overwintering sites for the shot hole fungus.

Symptoms and Damage. *Stigmina carpophila* causes shot hole spots, or lesions, on blossoms, leaves, fruits, and twigs. Only the current season's growth is susceptible to shot hole infection. Early in the season, small purplish areas develop on the young leaves. These areas expand into spots that range from 1/8 to 3/8 inch (3 to 10 mm) in diameter. In the early stages these lesions are surrounded by a narrow zone of light green to yellow colored tissue. Later on, leaf tissue in the lesions dies, turns light brown, and is surrounded by a dark margin. The dead tissue may fall out, leaving a "shot hole." Otherwise, the tissue may remain in the leaf and one or more small dark tufts form in the center of the lesion; these tufts are the fruiting structures of the shot hole fungus, and they produce spores. Some other agents, such as viruses, fungi, bacteria, and herbicides, may cause spots or shot holes, but none produce a fruiting structure in the lesion.

Infections that produce shot holes tend to occur on young leaves because young leaves expand as they grow, causing nongrowing dead tissue in lesions to break away and eventually fall out. Infected young leaves also drop more readily than mature diseased leaves, which may remain on the tree for the remainder of the growing season. Because mature leaves are no longer expanding, the dead lesion tissue is less likely to fall out. Lesions on mature leaves, therefore, are more likely to show fruiting structures and produce spores.

Small purple spots may also appear on the twigs. Examine these spots with a hand lens. If a black spore mass, or fruiting structure, is visible in the center of the lesion, it is shot hole.

On young fruit, the disease first appears as small, round, purplish spots which eventually develop a lighter center. These spots usually form on the upper side of the fruit, possibly because the hairs on the hull trap the water-borne spores. Many spots may run together. Fruit lesions become rough and corky as they grow older.

Shot hole can cause defoliation of trees, and, if left unchecked for several years, may debilitate trees and result in yield reductions. Severe fruit infections can cause fruit drop or gummy, deformed nutmeats.

Seasonal Development. During winter, the fungus survives within infected buds and twigs. Warm temperatures and free moisture are necessary for shot hole spores to be produced, germinate, and infect plant tissues. Spores are disseminated to leaves, twigs, and fruit in droplets of water. Persistent, warm rains during spring or sprinkler irrigation enhance spread and intensity of the disease, and lesions can appear 5 to 6 days after infection. The disease is found more frequently on lower parts of almond trees where sprinkler water reaches into the canopy. Lower limbs are gradually weakened by repeated loss of leaves and become unproductive.

Management Guidelines. Sprinkler irrigation intensifies disease incidence. In sprinkler-irrigated orchards where water hits the trees, you may reduce infection by shortening the irrigation period to reduce the time that trees are wet.

Contact fungicides provide control of shot hole if they cover the foliage and fruit during wet periods. The minimum number of applications required may vary each year depending upon the rain pattern and use of sprinkler irrigation. Make your control decisions based upon previous disease history, type of irrigation system you use, and expected weather patterns. Plan to apply at least one or two annual protectant sprays against shot hole. These sprays will also aid in the control of leaf blight.

In orchards with a high incidence of shot hole, make the application when leaf tissue emerges from the bud. This usually coincides with the pink bud to popcorn bloom stage, but may be closer to full bloom in some cultivars (see back cover). In orchards with a low incidence of shot hole, a single spray at the jacket stage of bloom can provide adequate disease control. Apply additional sprays, if rains continue, 1 to 2 weeks apart. Be sure to read the pesticide label and follow preharvest intervals.

Dormant application of a fungicide in November or early December should be considered only if the disease is so severe that twig infections are occurring.

Corky Spot

Causal Agent Unknown

Corky spot has been observed primarily in Contra Costa and Colusa counties. The cause of this disease is unknown, but it seems to be associated with local conditions because the disease is not perpetuated by propagation. Affected cultivars have been Nonpareil, Ne Plus Ultra, Long IXL, and Jordanolo.

Small, yellow necrotic spots appear on leaves, particularly around the margins. As the spots turn brown, they fall out giving the leaves a shot hole appearance. Premature leaf fall occurs, followed by new leaf formation. This pattern of shot holing, leaf drop, and new leaf formation may repeat itself several times during the season and can weaken the tree.

Corky spot lesions resemble those of shot hole, but are smaller than shot hole lesions and do not contain small fruiting bodies in the center of the lesion.

Corky spot has been associated with almond trees grown on almond rootstock rather than on peach rootstock. No other control measures are known.

Corky spot lesions may resemble those caused by shot hole, but generally lesions are smaller, concentrated around the leaf margin, and do not develop black spore structures in the center.

On twigs, scab lesions begin as water-soaked spots that eventually turn brown in the center and serve as overwintering sites for the fungus.

Scab infections first appear on leaves as small, yellowish spots.

Scab lesions eventually darken and look greasy.

Scab lesions on fruit tend to concentrate on the upper side of the hull and appear as small, dark, circular spots. Numerous lesions may coalesce to form large irregular, dark blotches, as seen on the nuts in the left side of this photo.

Almond Scab
Fusicladium (Cladosporium) carpophilum

Almond scab has been in California since the early part of this century. Previously it was most damaging in the northern Sacramento Valley, where it occasionally caused trees to defoliate. Scab seldom caused extensive damage until the 1950s, when use of sprinkler irrigation became widespread. Though all California almond cultivars are believed to be susceptible to scab, Ne Plus Ultra and Drake are most severely affected.

Scab infects leaves, fruit, and twigs in almonds, causing dark spots to form. The major concern with almond scab is partial or complete defoliation of the tree. If early defoliation is severe, fruit drop will also occur. Scab infections left uncontrolled for several years will weaken trees and reduce crop yield.

On twigs, lesions start as indistinct, water-soaked spots that gradually turn brown in the center and have a lighter colored margin. In spring, when the fungus resumes growth, it produces spores at the margin of the lesions, causing the margin to develop a distinct dark color. Lesions are superficial and do not girdle the shoots. Twig lesions are present throughout the year and can be used to confirm the presence of the disease in an orchard.

The first visible evidence of a scab infection on leaves occurs in June. Small, indistinct, yellowish spots develop on the under surface of infected leaves. These lesions gradually turn brown and become visible on the upper leaf surface. On leaf petioles and midribs, lesions are the same color but are more elongate.

Small circular spots can also develop on the hulls. These spots are more common on the upper side of the nut; eventually they coalesce into large, irregular dark blotches. The kernel is not affected, however. Fruit lesions caused by the fungus *Alternaria*, often referred to as black rot, may be confused with scab lesions because both appear on the upper fruit surfaces and often coalesce.

Scab lesions look greasy and oily and also occur on leaves and twigs. Black rot lesions occur only on the hulls.

The scab fungus overwinters in lesions on infected twigs and resumes growth in March. It is spread by spring rains and sprinkler irrigation, but leaf and fruit lesions usually do not appear until June or July. An annual treatment for shot hole should reduce scab.

Leaf Blight
Hendersonia rubi

Leaf blight of almonds was first identified in 1950, when it developed in a few Sacramento Valley orchards. Areas of high spring rainfall, such as Contra Costa County and the Sacramento Valley, have the highest incidence of leaf blight. Almond cultivars most susceptible to leaf blight are Drake, Ne Plus Ultra, and Peerless; moderately susceptible cultivars are Nonpareil, Mission, and IXL.

The most characteristic symptom of leaf blight is sudden withering of leaves. Leaves at the base of shoots and spurs begin to show symptoms in June; first they become yellow, then wither and die. Some affected leaves fall off, but most remain on the tree. Other leaves on the same spurs may wither and die throughout the summer. No sign of leaf blight fungus is present on affected leaves, although they may be covered with the dark growth of other fungi. (Photo on page 116.)

The fungus overwinters in the base of infected leaf petioles that remain on twigs throughout the winter. During late autumn and winter, the fungus may grow into the twig from the petiole and kill axillary buds. After the first fall rains, the lower end of the petiole turns light tan and becomes covered with dark fruiting bodies. Spores are spread by winter and spring rains, and the fungus infects petioles of new leaves in spring. Once inside the petiole, the fungus interferes with water conduction and the leaf withers. Flowers also may be destroyed in spring as they emerge from buds adjacent to infected leaf petioles.

The disease is not a serious problem unless it destroys leaves several years in a row and weakens the tree. Death of leaf and flower buds may contribute to yield loss. Leaf blight is controlled by the same practices and materials recommended for shot hole.

Green Fruit Rot
Botrytis cinerea and *Sclerotinia sclerotiorum*

Green fruit rot, also called jacket rot, calyx rot, and blossom rot, occurs on almonds, apricots, prunes, plums, and occasionally on cherry trees. The disease is of little importance in most years, but sometimes causes severe crop losses in all of the almond-growing regions. Ideal conditions for green fruit rot are cool temperatures and high levels of moisture during bloom. The disease is most damaging to cultivars such as Ne Plus Ultra that have tight clusters of blossoms.

In addition to *Botrytis cinerea* and *Sclerotinia sclerotiorum, Monilinia* spp. may also be involved in green fruit rot.

Symptoms and Damage. Green fruit rot begins during the latter part of the bloom period when the fungus infects senescing petals and anthers. Infected petals develop water-soaked brown spots. Usually diseased petals just fall to the ground, but some falling petals may land on leaves and cause secondary infections. Anther infections, on the other hand, spread to senescing floral tubes or flower jackets. Diseased jackets wither and stick to the developing fruit, especially if the jacket becomes wet. As the fruit sets and starts to grow, a brown spot develops where the jacket is adhered. The infected spot expands rapidly until the entire fruit rots. Frequently, fruit will fall off the tree, and the net result of the damage is a reduction in yield. (Photos on page 116.)

Seasonal Development. *Botrytis cinerea* overwinters in decaying plant material and possibly in broadleaf winter annual weeds such as filaree, shepherdspurse, and redmaids. It produces and releases spores in cool, wet weather. Spore release is greatest after a rapid rise or fall in relative humidity, but spore germination requires moisture. Spores are spread by air currents and deposited on blossom parts; they can survive 1 to 3 months if the weather remains cool and humid. Once an infection is established, the level of moisture and the temperature are less important for survival of the fungus.

Initially, blossom parts are somewhat resistant to infection by *Botrytis* spores, but as soon as they begin to wither, they become highly susceptible. Decaying flowers provide the necessary nutrients for spores to germinate and the fungus to grow. *Botrytis* spores easily penetrate withered flower parts. Infection of leaves or fruit requires contact with an infected flower part. The fungus can be identified by observing with a hand lens the typical clusters of spores produced on branching sporophores.

Sclerotinia sclerotiorum overwinters in the soil. In the spring during bloom, it produces spores that are carried to the blossoms by wind. If adequate moisture is present, the spores germinate and infect flower parts. The white fungal growth of *Sclerotinia* is readily distinguishable from the gray fungal growth of *Botrytis*.

Management Guidelines. Green fruit rot does not usually occur yearly. When bloom is extended by cool temperatures and moisture is abundant, apply a fungicidal spray at full bloom to prevent green fruit rot. Be sure the fungicide you use is effective against the two green fruit rot pathogens (*Botrytis cinerea and Sclerotinia sclero-*

tiorum), as well as *Monilinia* spp., which are primarily responsible for diseases such as brown rot and hull rot but also may be involved in the green fruit rot complex.

Yellow Bud Mosaic

A Strain of Tomato Ring Spot Virus

Yellow bud mosaic was first found in California in 1936 on peaches and almonds in Solano and Yolo counties. It affects many stone fruits, but in almonds, yellow bud mosaic seriously damages only the Mission cultivar. Many native plants provide natural reservoirs of the virus, which is transmitted from plant to plant by the dagger nematode.

Symptoms and Damage. Mission trees with yellow bud mosaic appear open and stunted; foliage is crinkled and distorted. Occasionally, chlorotic spots may develop that eventually die and drop out of the leaf causing holes or tattering. Leaves may also be stunted and form small rosettes or short tufts. They stand out in sharp contrast to normal appearing leaves on the same branch or in other parts of the tree. In severe cases the leaves may develop a mottled appearance.

Most of the leaves are concentrated at the branch terminals. Branches lack lateral growth and have little terminal growth from one year to the next. Frequently, one or more normal shoots are produced on a branch that is otherwise completely diseased. Fruits on diseased trees have wrinkled, rough hulls and appear larger than healthy fruit because the hulls are abnormally thick. Trees with yellow bud mosaic will live for many years but do not produce well; yield reduction is directly proportional to the severity of the symptoms.

The virus can also occur in other cultivars, but it rarely spreads up from the roots into the main portion of the tree, where most of the damage caused by this virus occurs. Occasional symptoms may be seen on water sprouts in the lower scaffold limbs of Nonpareil and Ne Plus Ultra, and a form of the disease has occurred in the Patterson area of Stanislaus County that caused Nonpareil trees to grow in a spiral form, similar to the characteristic spiraling of the IXL cultivar.

Seasonal Development. Yellow bud mosaic virus is indigenous to California and is widely distributed in the coastal areas and Sacramento Valley; it also occurs in a few scattered areas of the San Joaquin Valley. Many natural hosts of this virus (Table 14) show no symptoms of disease.

The virus is transmitted from infected host plants to almonds during root feeding by the dagger nematode, *Xiphinema americanum*. Because the virus enters through the roots, symptoms first appear on the lowest branches.

Table 14. A Partial List of Plants That Host Yellow Bud Mosaic Virus.

Common name	Scientific name
apricot	*Prunus armeniaca*
bristly oxtongue	*Picris echioides*
cheeseweed	*Malva parviflora*
common currant	*Ribes sativum*
Fullers teasel	*Disacus fullonum*
Himalaya blackberry	*Rubus procerus*
peach	*Prunus persica*
plantain	*Plantago* spp.
red raspberry	*Rubus idaeus*
sand strawberry	*Fragaria chiloensis*
seaside daisy	*Erigeron glaucus*

Gradually, over the course of several years, the virus spreads to the upper shoot terminals. If nematodes carrying the virus feed only on one root, the disease initially will develop only on one side of the tree. If the nematodes are distributed throughout the soil and feed on all the roots, the disease will progress uniformly throughout the tree.

Yellow bud mosaic virus is spread to new locations when floods or cultural operations move infected nematodes or portions of diseased trees. Trees in areas adjacent to east/west streams flowing from the coastal mountains or Sierra foothills have a higher incidence of this disease. Disease also could spread when seeds of infected wild plants germinate and grow in or nearby an orchard, and dagger nematodes are present in the soil.

Within an orchard, the disease spreads slowly to adjacent trees, gradually enlarging the original area of infection. Frequently, symptoms first develop in border trees and subsequently spread to adjacent trees. Although it is possible for infection to spread from tree to tree through root grafts, this is not common. The disease spreads mainly by the movement of dagger nematodes that carry the virus through the soil.

Mission trees planted in infested soil will exhibit symptoms in 2 years and will probably never be economically productive. When older Mission trees become infected as a result of natural spread of nematodes in the soil, they generally become uneconomical producers 3 to 5 years after the symptoms first appear.

Management Guidelines. Once yellow bud mosaic virus appears in an orchard, it is extremely difficult to eliminate. Deep soil fumigation at rates high enough to kill the nematode is expensive and often not practical, especially if other factors may also be affecting growth of the tree.

If you have an area that contains yellow bud mosaic virus, try to limit its spread to other parts of the orchard. Do not perform any cultural operations, including flood

Dead leaves at the base of this leaf cluster are typical of leaf blight infection. Next to the dead leaves, on the previous year's growth, are infected petioles remaining from leaves infected the previous season. These petioles serve as an overwintering site for the fungus.

Trees with yellow bud mosaic may have rosettes of stunted leaves occurring on branches with normal appearing leaves. The fruit on Mission trees appears wrinkled, rough, and larger than normal, because the hulls are abnormally thick.

Botrytis spores easily penetrate withered flower parts. Small nuts are susceptible to green fruit rot until the withered flower jacket falls off the nut.

Trees infected with yellow bud mosaic for several years have foliage that is concentrated on the terminals of branches that lack lateral growth.

Botrytis produces clusters of spores on diseased flower parts.

The brown area at the top of this nut indicates a green fruit rot infection. Eventually the infected spot will expand and the entire fruit will rot.

irrigation, that may move soil from this area to other locations. Remove infected trees and trees in at least two rows beyond, if the affected area involves only a small part of the orchard. If several areas of diseased trees are present, you may need to remove the entire block. Before removing the trees, treat them with an herbicide or girdle them to aid in killing the roots. Remove as many roots as possible. For best results, fallow the ground for 2 years and then fumigate the soil to kill the nematodes.

Do not plant Mission trees on peach or almond rootstock in infected areas. Marianna 2624 is resistant to the virus. If you do not fumigate, replant with cultivars that are compatible with Marianna 2624, or with non-hosts of the virus, such as walnut, pears, and most plums and prunes on Marianna 2624 rootstock.

Dieback of shoots and fruiting wood due to hull rot fungi is the most damaging aspect of this disease.

Hull Rot
Rhizopus stolonifer and *Monilinia* spp.

Hull rot can be caused by either the common bread mold, *Rhizopus stolonifer*, or by the brown rot fungi, *Monilinia fructicola* and *M. laxa*. Of the two *Monilinia* species associated with hull rot, *M. fructicola* is responsible for most hull rot in almonds. Although it is occasionally associated with hull rot, *M. laxa* is primarily responsible for causing brown rot blossom and twig blight in the spring.

Hull rot was first noticed causing severe twig blight in the 1950s and now occurs in most almond-growing districts. It is most severe, however, in well-managed orchards with large vigorous trees.

Hull rot caused by *Rhizopus* is evidenced by black fungi growing inside the hull.

Symptoms and Damage. The first indication of hull rot usually comes several weeks before harvest when leaves on a shoot wither and die. Closely examine fruit on this shoot for a brown area on the outside of the hull and either tan colored fungal growth in the brown area on the inside or outside of the hull (*Monilinia* spp.), or black colored fungal growth on the inside of the hull (*Rhizopus* spp.). Fungi invade almond hulls after the hulls begin to split in early July. Once in the hull, fungi apparently produce a toxin that causes the shoot and leaves to die from the point where the nut is attached to the tip of the spur. Fungi can also move in the other direction on the branch. Because the shoot is killed, other green nuts on the shoot won't mature, and they remain on the tree after harvest.

In sprinkler-irrigated orchards, hull rot frequently occurs on lower tree branches that are hit directly by water. Such infections can result in a loss of lower limbs on many of the trees throughout the orchard.

The most damaging aspect of this disease is dieback of shoots and fruiting wood and the resulting yield reduction. When more than 30% of the fruiting wood is killed,

Tan colored fungus growth on the outside of the hull indicates *Monilinia* hull rot.

a significant reduction in yield results. If less than 20% of the wood is killed, economic losses are minimized by the fact that most trees can offset the loss with new growth. Lower branches, however, do not usually generate new growth and will be lost.

Seasonal Development. Almond hulls are susceptible to hull rot fungi from the beginning of hull split until the hulls dry—a period that can range from 10 days to 2 months. Several factors affect the length of time it takes hulls to dry, including the amount of fertilization and irrigation an orchard receives, the amount of space between trees, and the openness of the tree's canopy. Rainfall or sprinkler irrigation between hull split and harvest will also slow drying and increase the length of time that hulls are susceptible to infection.

Rhizopus spores are commonly found in the soil. *Monilinia laxa* spores are found on blighted blossoms and sticktights on the tree. It is not known where spores of *M. fructicola* reside in the orchard. Spores of both *Rhizopus* and *Monilinia* fungi are readily airborne and can be spread by winds or by dried fruit beetles that are attracted to almond hulls when the hulls split. The first nuts on a tree to split are usually blanks, or nuts without kernels. Beetles are attracted to these hulls and bring fungal spores with them. Spores develop in blanks and provide a ready source of inoculum for the remainder of the nuts on the tree.

Trees with dense canopies are more severely affected by hull rot than trees with more open canopies, where air movement and greater degree of solar exposure reduce hull moisture and therefore the hull split period. Because of increased sun exposure, trees on the border of the orchard have a lower incidence of hull rot than trees within.

Management Guidelines. When planting an orchard, consider selecting resistant cultivars. The most susceptible cultivars are Nonpareil, Jordanolo, and IXL. Merced, Thompson, and Ne Plus Ultra are frequently affected by the disease, but the damage is minimal. Hardshell cultivars such as Mission, Davey, and Drake may exhibit rotted hulls, but shoot dieback is rare in these cultivars.

To reduce the incidence of hull rot, follow cultural practices that increase uniformity of hull split and decrease drying time of the hulls without sacrificing yield or kernel quality. For example, prune trees to increase the amount of sun exposure that the almonds receive, but not to the point where yield is sacrificed. This will improve uniformity of hull split.

Not irrigating the orchard during the period from hull split to harvest may also help lower the incidence of hull rot. Drying the soil in the orchard will decrease the length of time it takes for hulls to dry and may reduce the number of beetles emerging from the orchard, thus reducing spread of the fungi.

Harvest your crop as soon as nuts at eye level show 95 to 100% hull split. Early harvest will lower the incidence of twig dieback. Fungi remain in the hulls, and once nuts are removed, twig dieback ceases. Shake your trees twice, if necessary, to remove later maturing nuts.

After harvest, removing any infected nuts remaining on the tree may help to reduce the level of inoculum in the orchard. It will also serve to eliminate overwintering sites for the navel orangeworm.

Spring fungicide treatments to control brown rot blossom and twig blight will not effectively control hull rot in the fall. In general, fungicide treatments are not effective against this disease.

BUD FAILURE DISORDERS

Bud failure disorders can affect primarily either vegetative buds (noninfectious bud failure) or flower buds (nonproductivity). Flower bud failure results from a loss of flower buds or a failure of the flowers to set and may be associated with a virus (as in almond calico) or genetically induced (as in nonproductive syndrome).

Noninfectious Bud Failure

Noninfectious bud failure, also known as crazytop and muletail, is classed as "genetic," because the disorder has not been transmitted from an infected plant to a healthy one. It is perpetuated by vegetative propagation. Seedlings can inherit it from their parents, but the exact nature of the causal factor is not known.

The disorder is associated with specific cultivars. Before 1940, bud failure was observed sporadically in Nonpareil and Peerless, but the problem became acute when the Jordanolo cultivar was released in 1937. Jordanolo is an offspring of Nonpareil X Harriott. Since that time, noninfectious bud failure has appeared in other new cultivars including Jubilee, Merced, Harpareil, Thompson, Harvey, Arboleda, and Carmel. A few affected Mission trees also have been observed.

The disorder occurs to some extent throughout all almond-growing regions in California, but is most severe in the southern San Joaquin Valley and western and northern Sacramento Valley, where high summer temperatures accentuate its occurrence. Incidence and severity of bud failure within a given orchard is variable and can involve parts of a tree, a few random trees, groups of trees, or all the trees of a certain cultivar.

Symptoms and Damage. The most conspicuous symptom of noninfectious bud failure is failure of vegetative buds to grow in the spring. If you examine individual vegetative or shoot buds during fall or winter, they often will appear necrotic. Flower buds are not affected by this disorder once they have developed. If symptoms are severe, however, fewer flower buds will develop because vegetative growth is reduced (fruit bearing wood), and flowering time may be delayed. Percent fruit set is generally good, considering the fewer number of blossoms produced.

The percentage of vegetative buds on a shoot that fails determines the severity of the symptoms. Buds are often viable at the base of the shoot, nonviable in the middle, and occasionally viable at the tip of the shoot.

The following spring, viable buds grow into shoots, often at right angles to the branch, and are unusually vigorous. Look for bud failure symptoms just after growth begins in spring and for the next 6 to 8 weeks (through April). Examine previous years' shoots for necrotic, failed buds (some of these buds may have already abscised and dropped). Later in the season new shoot growth may obscure symptoms and make it difficult to distinguish between vegetative bud drop caused by noninfectious bud failure and bud failure from other causes.

Trees that have shown bud failure symptoms for several years have sparse foliage because fewer shoots are produced, and no leaves occur at the nodes of older twigs. Unusually large numbers of branches are clustered at the base or tip of lateral branches. Often branches have changed their direction of growth several times—some grow back upon themselves. This growth behavior and the right-angled shoots give the branch system a crooked, tangled appearance. Names such as *crazytop* refer to this condition.

Rough bark is another symptom associated with noninfectious bud failure, but its occurrence is sporadic. This symptom occurs when necrosis from the shoot bud extends to the surrounding stem tissue. It shows up on new shoots during summer or fall, often starting as a discolored area at the base of a leaf. As the shoots get older, these areas continue to develop, expand, and become more prominent. Rough bark frequently occurs on the middle or upper portions of shoots in 5- to 6-inch (12.5 to 15 cm) bands along the branch. Viable buds do not form in these areas and symptoms can be quite dramatic on young trees. Entire shoots may be devoid of viable lateral buds, and few normal buds will develop at the shoot tips.

The effect of bud failure on yield depends on severity of the disorder in a particular tree. Symptoms may increase with age, but usually yield reductions reach a plateau in most cultivars; yields may fluctuate about this plateau, depending on seasonal temperatures, but generally are stabilized.

Seasonal Development. Vegetative bud necrosis and rough bark develop during midsummer and early fall after exposure to periods of high temperatures. Lack of moisture and other stresses also may accentuate the disorder, but the relationship between temperature and moisture stress is not clear. Variation in severity of symptoms from year to year is related to the temperature pattern during the previous growing season.

Symptoms in affected trees can appear at any age from the first growing season to more than 10 years later, but most trees develop symptoms when they are about 4 to 5 years old. Some trees do not develop symptoms, however, until after 15 to 20 years.

The age at which a tree expresses bud failure symptoms and the severity of these symptoms depend on 1) the relative bud failure potential in the source plant from which buds used for propagation were taken, and 2) the environmental conditions to which the tree is exposed. Trees with bud failure develop symptoms at an earlier age and to a greater degree in the hotter areas of the southern San Joaquin Valley than in the cooler summer areas of central California and the coastal valleys. Once bud failure symptoms begin to appear in a tree, the disorder will remain. Although developing buds are killed by high temperatures in July and August, they don't drop off until the following spring.

Management Guidelines. The only known cure for bud failure is to remove or top-work affected trees. Trees that aren't severely affected by bud failure can be left in the orchard. Their crop potential may be lower than normal trees, but they will not die from the disorder, and the disorder will not spread from tree to tree.

Usually it is not economical to replace bud failure trees unless they are under 10 years old and have greater than 25 to 30% reduction in yield. Young trees that show symptoms in their first or second year or older trees that are severely affected should probably be removed and replanted.

Severely affected trees less than 10 years old also can be top-worked to a different cultivar that is free from bud failure or to the same cultivar. If you use the same cultivar, use propagation material from a source that has some record of bud failure resistance. Although top-working requires skill and postgrafting care to be successful, it is the fastest method to bring a tree back into production.

Selecting a source of suitable scion wood for top-working is difficult because the potential for bud failure cannot be identified by visual inspection unless the tree actually shows symptoms. It is important, however, to inspect a bud-source orchard for presence of affected trees. If bud failure trees occur, avoid the entire orchard. It is also important to select a source that has a known record of propagations through commercial use or has been progeny tested and found relatively free of bud failure. Lacking this information, it is best to select source orchards located in the cooler, central portion of the Central Valley rather than in the northern Sacramento or southern San Joaquin valleys because of the association of bud failure development over time with high temperature exposure.

If you use cultivars that are known to be affected by this disorder, it is difficult to completely avoid the use of some trees with bud failure susceptibility, regardless of the source of the vegetative material.

A typical overall symptom of noninfectious bud failure is sparse foliage concentrated on new growth.

This shoot illustrates the pattern of growth associated with noninfectious bud failure that is known as crazytop.

Crazytop caused by noninfectious bud failure is easiest to see during dormancy when the leaves are off the tree.

Almond Calico

A Strain of *Prunus* Ring Spot Virus

Almond calico occurs throughout the almond-growing regions in California. The almond calico strain of virus belongs to a group of related viruses called *Prunus* ring spot virus. *Prunus* ring spot virus has many strains, and there is great variability among them. Variants of strains can occur in the same tree and, by interfering with each other, markedly affect the distribution of symptoms within a particular tree. Thus, some strains may not particularly affect tree productivity; others can greatly reduce productivity. Likewise, different cultivars can react differently to a particular strain. In addition to infecting almonds, the almond calico strain of this virus also infects other stone fruits. In peach it causes the disease known as peach mule ear.

Cross sections of vegetative buds from trees with noninfectious bud failure frequently reveal brown areas of dead tissue.

The band of rough, cracked bark located to the left of the nut cluster is one of the symptoms associated with severe noninfectious bud failure.

Symptoms and Damage. Almond calico symptoms first appear in most cultivars as a mottling of the leaves on infected branches. In subsequent years, leaves produced in the early part of the season develop chlorotic rings or bands that gradually become creamy white. Characteristic patterns include lines, rings, and chevrons. Some of these leaves may drop prematurely, but leaf symptoms do not normally cause economic losses.

Leaf symptoms may not always occur yearly in infected trees, but usually are accentuated in years with cool spring temperatures. A necrotic strain of *Prunus* ring spot virus also infects almonds and produces somewhat similar leaf symptoms, but usually these are not chronic. If leaf symptoms are caused by the necrotic strain, they will gradually become milder, and the tree will recover and not develop bud failure symptoms.

Moderate to severe flower bud failure frequently develops in trees affected by almond calico. This may not occur until 10 years or more following the first appearance of leaf symptoms. Lateral flower buds on shoots produced the previous season fail to bloom, or those that do, produce few flowers that set. Likewise, lateral leaf buds and even terminal shoot buds may die. Bare shoots occur because vegetative buds are either not produced in the leaf axil ("blind nodes"), or if they are, the bud dies and falls off before leafing out. If vegetative buds do not grow, the shoot will die back to the uppermost live lateral vegetative bud. If most of the lateral vegetative buds die and terminal buds continue to grow, the shoot becomes long and willowy with leaves and fruit only present on the terminal portion. As trees age, flower and vegetative bud failure symptoms become more prominent and result in an overall reduction in yield.

To distinguish bud failure associated with almond calico from genetic bud failure, look for such leaf symptoms as mottled leaves with chlorotic blotches. Trees with noninfectious, or genetic bud failure, do not have leaf symptoms unless they are also infected with almond calico virus. Also, genetic bud failure affects primarily leaf buds and not flower buds. Almond calico bud failure affects both leaf and flower buds.

Trees of the Drake cultivar do not exhibit prominent calico leaf symptoms, but they do develop flower and leaf bud failure. Affected Mission trees commonly have no calico leaf symptoms or bud failure, but may fail to set fruit several years following infection, and tend to have leaves that are longer in proportion to width. These symptoms resemble those caused by nonproductive syndrome. Because Mission trees do not exhibit diagnostic leaf and bud failure symptoms, almond calico may be difficult to identify in this cultivar unless indexing is performed to determine if the almond calico virus is present.

Seasonal Development. Almond calico virus is thought to be transmitted in pollen that is produced in flowers on infected branches. Pollen may infect other trees during bloom when bees carry infected pollen to healthy trees. Leaf symptoms may or may not occur during the first

Trees with almond calico have bare shoots with thin foliation and few fruits. Almond cultivars that grow upright and erect resemble the tree in this photo when they are infected.

Calico symptoms in almond cultivars with spreading growth frequently cause the tree to appear "brushy."

Leaves with almond calico frequently have chlorotic bands.

year of infection. In the second and third year, however, leaves on affected branches show line patterns and chlorotic mottling. Although this disease does not kill the tree, leaf symptoms recur throughout the tree's life, and flower and vegetative bud failure symptoms may gradually intensify as trees age.

Management Guidelines. Both the scion cultivar and the rootstock of nursery trees can be infected with almond calico virus. This disease can be avoided by obtaining trees that have been propagated from source trees tested free of *Prunus* ring spot virus planted on rootstocks that come from ring-spot-free seed trees. This can be accomplished by buying certified nursery trees, if they are available.

If a few isolated trees in your orchard have almond calico, consider removing them. Their pollen may serve as a source of inoculum for other trees during bloom.

If the disease is widespread in your orchard, you may eventually want to replant the entire orchard if and when production appears to be affected. This decision should be based on the overall productivity and profitability of the trees. Do not replant a small portion of the orchard if the disease is widespread. When newly planted trees come into bearing, they are likely to be infected by pollen from older infected trees.

Nonproductive Syndrome

Large, vigorous, upright trees with nonproductive syndrome are often referred to as "bull trees" or "rooster trees." The terms "bull" and "rooster" are general ones, however, and have also been used in reference to trees affected with virus-induced bud failure. Trees with nonproductive syndrome have varying production ranging from very low to near normal, distorted nuts with softer than normal shells, and long, somewhat narrow leaves. In recent years, this condition has been most commonly observed in the Mission cultivar, where it is referred to as "bull mission," but a similar condition also occurs in Nonpareil, Carmel, and Fritz. (Photos on page 124.)

In Mission, severity of symptoms may vary greatly from one tree to the next, and even from one branch to the next within the same tree. Productivity, as measured by percent fruit set, can range from zero to near normal levels (30 to 50% set). Nuts on affected trees tend to have distorted shapes; they might be longer and narrower than normal nuts, or they can be shorter and plumper. Frequently, nuts on affected Mission trees have soft shells with a prominent wing extending from the suture, instead of the hard shell that is characteristic of that cultivar.

Nonproductivity results from a range of abnormalities in the reproductive system, including defective pistils, ovules, and pollen grains—and flower and nut drop. Not all of these abnormalities occur on every tree that is affected, but their prominence increases with the increasing severity of the condition. These abnormalities sometimes occur simultaneously with other variations, such as bitter kernels.

In addition to reproductive abnormalities, crinkle-leaved or rough bark, nonproductive *sports* (single branches that are characteristically different from other branches on the same tree) are occasionally observed in single trees in orchards.

Symptoms associated with nonproductive syndrome appear to be the result of genetic changes, or mutations, within cells. Mutant cells may occur in the growing points (buds) of trees, but their presence may not be detected until a large proportion of the cells in the branch are affected. If the mutations that cause nonproductive syndrome occurred in the plant used for propagation, offspring trees may be produced with the disorder, depending on the proportion of affected cells in the bud used. Affected trees in an orchard often occur in groups of approximately ten trees, which can be associated with nursery bundles of ten. In nurseries, trees from a specific bud source are frequently grouped together. In addition, affected trees are unusually vigorous, which increases the likelihood of being grouped together during size grading.

Because other conditions such as young age, high vigor, some mineral deficiencies, lack of pollination, or specific diseases can reduce percent fruit set and modify nut characters, careful diagnosis of nonproductive syndrome is important. Shot hole fungus, for instance, can reduce crops and produce distorted nuts. Some virus diseases, such as almond calico, are also associated with nonproductivity and can resemble this condition to a certain degree. Corroborative indexing, a method of identifying viruses (see *indexing* in glossary), can be used to distinguish between viral and nonviral conditions.

If trees with nonproductive syndrome appear in an orchard, replace the ones with significantly reduced yields either by replanting or by grafting. Selection of nonaffected source trees for propagation material is essential but difficult because lack of visual symptoms on the bud source tree does not prove absence of nonproductive syndrome.

ENVIRONMENTALLY-CAUSED DISEASES

Pathogens and genetic disorders may not be the only causes of disease symptoms on leaves and nuts. Adverse weather, soil conditions, nutritional disorders and the like produce symptoms as well.

Gumming, "Split-Pit," and Corky Growth in Nuts and Kernels

Occasionally, almond nuts show protrusions of gum during development. You may find gum pockets, cracked shells, and aborted, gummy, or poorly developed kernels. In nuts of the Jordanolo cultivar, a thin layer of corky material may be stuck to the kernel. These conditions are associated with larger than normal nut sizes and generally are the result of a rapid growth rate during early stages of fruit development in April and May. The innermost layer of the shell hardens and stops growing. It eventually cracks because the outer layers of the shell are still growing. Gum is produced within the fruit as a result of this injury. Callus tissue may develop from cracks and grow next to the kernel to form corky material that dries and sticks to the kernel when mature. (Photos on page 124.)

Fruits tend to grow more rapidly in the early stages of development if certain growth conditions, such as a light crop and early season moisture stress followed by irrigation, occur within the orchard. Thompson and Jordanolo are more subject to corky growth, gumming, and split-pit than other almond cultivars. To avoid these problems, do not plant Thompson and Jordanolo cultivars, and correct any orchard condition that contributes to excessively vigorous growth and light crops.

Nutritional Disorders

An excess or deficiency of minerals in the soil is harmful to the tree because it interferes with its growth and development. Symptoms produced by trees with nutritional disorders may be confused with symptoms caused by other stresses such as diseases, herbicide toxicities, and environmental pollutants. To identify nutritional deficiences or toxicities, evaluate leaf symptoms in conjunction with a leaf tissue or soil (toxicities only) analysis—or both. (See section on diagnosing mineral deficiencies and toxicities in *Almond Orchard Management*, listed under General References.) (Photos on page 124.)

Toxicities. Nutrient toxicities most likely to occur in California almond orchards are excesses of boron, chlorine, and sodium. Toxic levels of these minerals may be present in the soil or build up because of poor quality irrigation water. Symptoms caused by these toxicities are most evident from July through September.

Boron. Boron toxicity and damage is frequently encountered on the west side of the San Joaquin Valley as new plantings expand into soils high in boron. Toxicity causes shoots to die back at the tip and the bark to become cracked and corky. Midribs on the undersides of leaves also develop corky areas.

Chloride and Sodium Toxicity. Toxicity caused by excess levels of chloride and sodium occurs in limited sites scattered throughout all almond-producing districts. Both produce leaf-tip burn symptoms. Use a leaf analysis to distinguish between sodium and chloride toxicities.

Deficiencies. The most common nutrient deficiencies in California almonds are, in order of importance: nitrogen, zinc, potassium, boron, and iron. Copper, magnesium, and manganese deficiencies can also occur, but these are rare.

Nitrogen. Almost all orchards will become nitrogen deficient if nitrogen is not applied. Nitrogen deficiency in a block of trees is difficult to diagnose by symptoms alone because all the trees in the block may look the same. Have a leaf analysis performed to confirm nitrogen deficiency.

Early in the season, trees may show reduced shoot growth and leaves may be small and pale. Foliage also may be sparse. In autumn the leaves become yellow and drop earlier than normal. Insufficient or excess soil moisture may produce similar symptoms.

Zinc. Zinc deficiencies are common in the sandy soils of the San Joaquin and Sacramento valleys, but can appear on trees grown in heavier soils as well. Symptoms appear early in the season, especially if the deficiency is severe. Zinc deficient trees bloom and leaf out later than normal trees; in some cases they may be as much as a month behind in development. When leaf buds do open, leaves are small, chlorotic, and appear in tufts—giving rise to the term "little-leaf." If the deficiency is not so severe, leaves are only slightly smaller than normal but show chlorotic areas between the lateral veins. In severely deficient cases, terminal dieback sometimes occurs. As the season progresses, subsequent shoot growth tends to mask zinc deficiency symptoms that were present earlier, especially on older trees. Rapidly growing young trees may show symptoms throughout the season. Nuts on zinc deficient trees are smaller than normal nuts.

Potassium. Potassium deficiences can be found in the northern Sacramento Valley and in sandy areas of the San Joaquin Valley. Leaf symptoms appear in early to mid-summer; leaves become pale, resembling nitrogen deficiency, and leaf size and shoot growth are reduced. The tip and upper margins of leaves on vigorously growing shoots become necrotic. Leaf tips curl upward, causing a condition referred to as "Viking's prow." Overall nut size is reduced.

Boron. Boron deficiency is limited to a few areas in the northern San Joaquin Valley and the north and west sides of the Sacramento Valley. Terminal and subterminal leaves on vigorously growing shoots develop tip scorch, and eventually the entire leaf may become necrotic and fall off. Frequently the tips of these shoots die back. In the following year, several buds below the dieback produce shoots, resulting in a brushy appearance. Brown, gummy areas develop in the green nuts, and gum appears on the surface of the nut. These nuts often drop in May or June and can result in loss of the entire crop.

Nuts on Mission trees with nonreproductive syndrome tend to have distorted shapes (left); a normal nut on a healthy Mission tree (right).

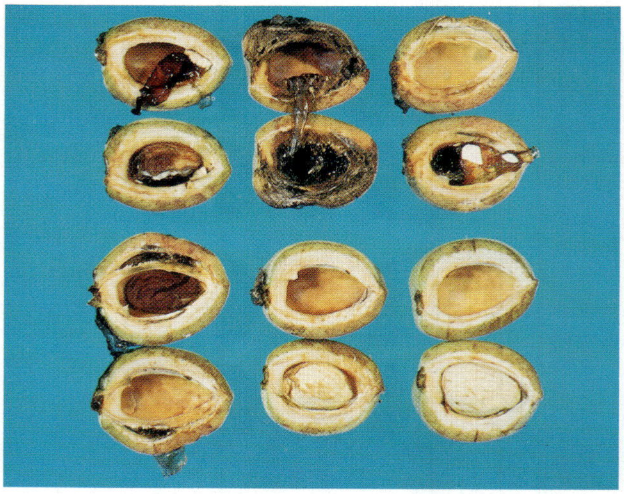

DON EDWARDS

Gum pockets, cracked shells, and aborted, gummy, or poorly developed kernels may all result from rapid growth during early stages of fruit development in April and May, especially in Thompson and Jordanolo cultivars.

Trees deficient in zinc have small, chlorotic leaves that appear in tufts.

Excess sodium produces tip burn on leaves of almond trees. Identical leaf symptoms may also be produced by chloride toxicity.

Potassium deficiency results in pale leaves with necrotic tips and edges. Frequently the leaf tips curl upwards creating the symptom known as "Viking's prow."

K. URIU

Almond trees deficient in boron may exhibit shoot dieback.

Iron deficiencies generally result in leaf chlorosis in the early season.

Iron. Iron chlorosis is occasionally found in the Paso Robles area and in a few scattered Central Valley locations. Chlorotic leaves are produced early in the season and the entire leaf is uniformly yellow. Chlorosis may disappear as the season progresses, or in severe cases, leaves may develop scorch and drop. Iron chlorosis symptoms are associated with calcareous soils or heavy, poorly drained soils.

Herbicide Symptoms

Tree injury is a risk if herbicides are misused and may occur under some conditions despite their proper use. Dosage rate, organic matter content, rainfall or irrigation, careless application, soil texture and depth of soil all bear on potential tree injury by herbicides. Tree injury can range from severe damage to slight effects on growth and leaf color that do not affect yield or quality.

Most crops generally display specific symptoms from a particular herbicide. Familiarity with these symptoms and with the history of herbicide use in and around the field can help you diagnose herbicide injury.

The almond leaf on the left is healthy; the others have been damaged by paraquat drift.

Leaf symptoms produced by dichlobenil on these prune leaves are similar to those that occur on almonds: leaf tip chlorosis that may spread around the entire leaf margin and cause interveinal chlorosis, depending on herbicide dosage.

Almond trees accidentally sprayed with glyphosate in the fall produce small, puckered leaves the following spring.

Simazine symptoms on almonds include marginal or interveinal chlorosis.

Norflurazon injury to almonds causes veinal chlorosis in the leaves.

ART LANGE

Dalapon injury results in marginal necrosis of mature leaves.

Paraquat. Paraquat drift damage on almond leaves appears as small, round, yellow spots that may turn brown but do not fall out, as they do in shot hole. Also, no small spot, or fruiting structure, develops in the center of the lesion as with a shot hole infection. If heavy concentrations are sprayed on leaves, it will burn the foliage, but new growth is normal. Tree trunks of newly planted trees can show brown spotting if the bark is still green when sprayed.

Dichlobenil (Casoron, Norosac). Leaf tips of mature leaves damaged by dichlobenil become chlorotic and burnt. Depending on the herbicide dosage, this chlorosis may spread around the entire leaf margin. Interveinal chlorosis and small burned areas may also develop. Symptoms usually appear in early to midsummer and become progressively more severe through the season.

Dalapon. Mature leaves injured by dalapon develop marginal necrosis that gradually moves into the leaf blade. Blotches of dead tissue, similar in appearance to sodium and chloride burn injury, develop, and if the leaves are still developing the leaf cups upward.

Glyphosate (Roundup). Two different types of damage symptoms may be caused by glyphosate depending on the time of year it is applied. If glyphosate is applied in the fall and gets on the trees, symptoms appear the following spring on new growth. Affected trees are slow to leaf out in spring and produce many small, puckered, almost needlelike leaves. Tree growth is slow throughout the season until the effect of the herbicide wears off. When glyphosate is applied in the spring and gets on the trees, young terminal leaves on the tree may turn yellow. High leaf concentration of the herbicide will cause these leaves to die. New growth that follows may appear yellow or blotchy yellow, or it may be completely normal.

Simazine (Princep). Simazine symptoms are frequently seen in almonds, especially in trees of the Mission and Merced cultivars grown in sandy soils that are highly alkaline. A slight overdose, sometimes due to failure to turn off the sprayer on turns, can give trees the characteristic ivory yellow marginal or interveinal chlorosis on mature leaves. Chlorosis starts on the leaf margins and moves toward the center, leaving the mid-vein and some of the lateral leaf veins green. High concentrations also may cause marginal burn. Symptoms usually appear in the early to midsummer and become more severe during periods of high temperatures.

Norflurazon (Solicam). Injury by norflurazon causes veinal chlorosis of leaves and a pinkish, light purple to completely white color on new shoot growth.

2, 4–D or MCPA. The phenoxy herbicides usually cause young limbs and leaf petioles to grow in a twisted manner. Symptoms are most frequently observed in early to midsummer. High concentrations of these chemicals may scorch leaves. New growth is also affected, but the symptoms gradually disappear.

Frost Damage

Dormant almond flower buds in midwinter are relatively insensitive to frost damage down to 20°F (−7°C). Sensitivity to frost injury increases gradually as the buds come out of dormancy. When the tree blooms, sensitivity increases tremendously. From mid-bloom until the fruits are the size of a pea or larger, the crop can be significantly damaged by temperatures only slightly below freezing. Spring frost can blacken and kill budding leaves, flowers, and young fruits.

The ability of almond flowers and nuts to withstand freezing temperatures may be related to the presence and density of certain bacterial species commonly found on plant surfaces. These bacteria are believed to serve as nuclei around which ice crystals form. If the bacteria are absent or inactivated, almond blossoms and fruit can resist ice formation down to temperatures as low as 24°F (−4.4°C). The most common ice-nucleating bacterium found on almonds is *Pseudomonas syringae*. This is the same bacterium responsible for causing bacterial blast, a disease associated with freezing temperatures, and bacterial canker.

Wind Injury

Proper pruning and training of young trees will help prevent wind damage. Insect pests and disease can render a tree susceptible to wind damage. Scaffold crotches may be weakened by wood-boring insects, and strong winds will cause the scaffolds to break. Wood-rotting fungi, like Poria wood rot, weaken older trees by causing the structural wood to rot. These trees may appear to be healthy but can be blown over by strong winds.

Sunburn Damage

The bark of young tree trunks and exposed limbs on older trees may be sensitive to sunburn. Sunburned bark is cracked, somewhat sunken and is susceptible to attack by wood borers and wood-rot fungi. Use tree wrap or paint trunks of young trees with white wash or a white, interior water-based paint to protect them against sunburn.

Nematodes

Nematodes are tiny, round worms that live in diverse habitats including soil, water, and plant tissues. Many damage plants, but others feed on fungi, bacteria, or other nematodes. Typically, both plant-feeding and non-plant-feeding nematodes inhabit orchard soils.

Nematode pests of almond trees are widely distributed throughout the principal fruit and nut producing areas of California, but damage to trees is more prevalent in sandy soils or soils with a sandy subsurface. The impact of nematode feeding is generally greater on trees with restricted roots systems.

The major nematode pests of almond trees and other deciduous fruit crops are root knot nematodes, *Meloidogyne* spp.; root lesion nematode, *Pratylenchus vulnus*; ring nematode, *Macroposthonia xenoplax*; and dagger nematode, *Xiphinema americanum*. Other plant-feeding nematodes such as pin nematode, *Paratylenchus hamatus*, and stubby root nematode, *Paratrichodorus christei*, are found in orchard soils, but more research is needed to determine if they cause economic damage.

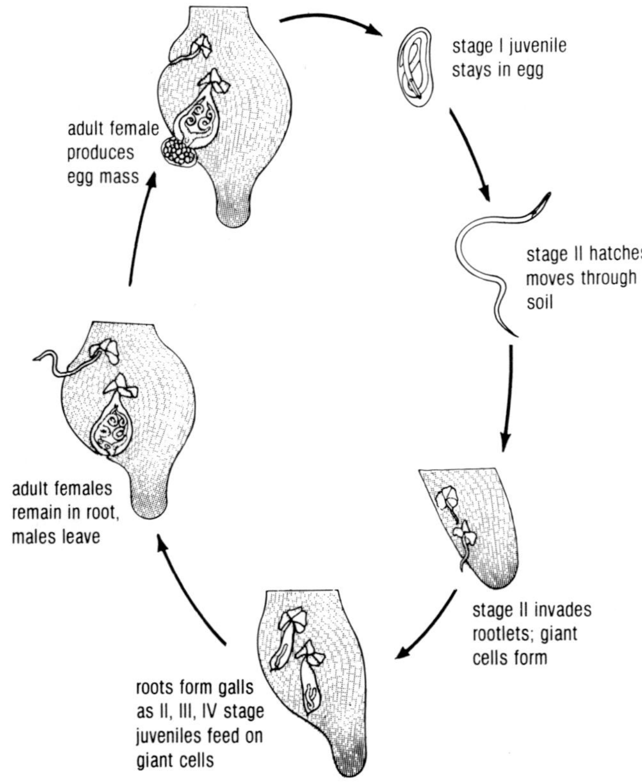

Figure 53. Root knot nematodes spend most of their life cycle in galls on roots. Second-stage larvae invade new sites, usually near root tips, causing some root cells to grow into giant cells, where nematodes feed. As feeding continues, the plant produces a gall around the infected area.

Nematode Species

Root Knot Nematode *Meloidogyne* spp.

At least four species of root knot nematodes occur in the almond-growing regions of the state. *Meloidogyne incognita* is the most common and widespread root knot species. *Meloidogyne hapla* occurs mainly in the northern growing regions or in association with alfalfa production. Two species, M. *javanica* and M. *arenaria*, occur sporadically throughout almond-growing regions. Because they all have similar life histories and are controlled in the same manner, there is currently no practical reason to distinguish the four species.

Eggs of the root knot nematode are deposited in a gelatinous mass in the root or on the root surface. Microscopic, second-stage larvae hatch from the eggs and enter roots just behind the root tip or at the junction where the roots branch (Figure 53). After entering the root, they establish themselves and begin to feed in a single location in the conductive tissues. As nematodes feed, the host root produces a gall around each nematode that gradually enlarges. The adult male remains wormlike, does not feed, and leaves the roots following its final molt. The female remains in the root and eventually swells into a flask shape. Often several females will inhabit a single gall.

Populations of root knot nematodes are most heavily concentrated in a zone between 6 to 36 inches (15 to 91 cm) beneath the soil surface, but can occur wherever roots occur. Before entering the root, second stage larvae are mobile and can move up, down, or laterally in soil.

Root knot nematodes disrupt root functions by stimulating the production of galls, which hinders flow of nutrients and water throughout the tree. If a tree is otherwise stressed, by lack of water, for example, damage caused by root knot nematode is more severe. When you suspect root knot nematode damage, look for the obvious but small galls on the roots. Damage by nematodes other than root knot is not obvious, so even if you do not find root galling in your orchard, get a soil sample diagnosis to confirm the types of nematodes present.

In California, root knot nematodes are most prevalent and damaging in coarse-textured soils including sand, loamy sand, and sandy loam. Even trees planted on root knot resistant rootstocks such as Nemaguard can be initially damaged if high populations are present when the trees are planted.

Root Lesion Nematode *Pratylenchus vulnus*

Several species of root lesion nematode can be found in orchard soils, but *Pratylenchus vulnus* is the only one known to damage almond trees. A second species of root lesion nematode, *P. brachyurus*, is occasionally found in southern San Joaquin Valley almond orchards, but the degree of damage it causes is unknown. *Pratylenchus minyus* is commonly present in grassy orchards but is not considered a pest of almond rootstocks.

Root lesion nematodes are migratory nematodes that live and feed either inside the root or in the soil outside the root when they are seeking new roots to penetrate. Females lay eggs within the roots or in the soil, and second, third, fourth and adult stages may feed on roots. If conditions in the root become unfavorable, juveniles or adults may leave the root and invade other nearby roots.

Root lesion nematodes girdle and effectively prune the roots in the process of feeding; small feeder roots are most susceptible. Lack of feeder roots on affected trees is a distinctive symptom of root lesion nematode feeding, but a soil analysis is necessary to confirm the root lesion nematode's presence. In some crops, such as walnuts, they cause a dark colored area, or lesion, on the root surface. On rootstocks used for almonds, however, lesions may not develop, so this symptom is not a good diagnostic characteristic.

Root lesion nematodes are most damaging in replant situations where soils have not been preplant fumigated, or in established plantings where root development has been restricted by a sandy subsurface, herbicide damage, shallow hardpan, poor land preparation or the like. Other hosts of this nematode are grape, walnut, fig, plum, prune, rose, cherry, and peach.

Ring and Dagger Nematodes *Macroposthonia xenoplax* and *Xiphinema americanum*

Ring and dagger nematodes do not enter tree roots but feed on them from the outside, mainly at the root tip. Both nematodes also play roles in the development and spread of certain diseases. Dagger nematodes transmit yellow bud mosaic virus to almond trees. Almond trees grown in soils containing high populations of ring nematode tend to be more susceptible to bacterial canker than almond trees grown in soils where ring nematodes are absent or occur in low numbers.

Ring nematodes can build up to large numbers in orchards or vineyards with coarse, sandy soils, or with fine-textured clay soils. They are most commonly found in orchards located in Merced and Stanislaus counties and the southern half of San Joaquin County. Dagger nematodes do well in many different soil textures.

Damage

Nematodes puncture plant cells with a needlelike device, called a stylet or spear, which enables the nematode to remove the cell contents. Nematodes are most damaging when trees are undergoing their major period of root development during the nonbearing years following planting. The first year is particularly critical; at this time high numbers of nematodes can hinder root development, even in trees that are planted on nematode-resistant rootstocks. After the first few years, young trees planted on root knot resistant rootstocks can tolerate feeding by root knot nematodes.

Vigorous, mature almond trees can tolerate feeding by low populations of root knot, root lesion or ring nematodes, but if populations get very high, even well-established trees can decline in vigor and yield. If you try to replant in a heavily infested orchard site, young tree roots can rapidly become heavily parasitized and produce little or no growth.

Nematode-infested trees have a poorly developed, inefficient root system and frequently exhibit nitrogen and other nutrient and water deficiencies. In addition to reducing root efficiency, nematode feeding is also associated with a reduction in fruit size. Nematode damage to the root creates a "sink effect," in which plant photosynthates, or nutrients, are diverted from the upper portion of the tree to the roots, making the nutrients unavailable for fruit production. Aboveground symptoms of nematode damage—lack of vigor, small leaves, dieback of twigs, and yield reduction—are poor diagnostic symptoms because these conditions can also be caused by many other factors. Except for the galls resulting from the presence of root knot nematode, the damage caused by nematodes is mostly nonspecific and difficult to diagnose. You must take soil and root samples and have nematode populations assessed by a laboratory in order to confirm that nematodes are limiting the growth of your trees.

Management Guidelines

Most nematode research in stone fruits has been conducted on peaches, nectarines, or plum, rather than on almonds. The management guidelines presented here are mostly derived from experience with these crops. Peach, nectarine, and plum trees are planted on the same rootstocks as almonds, but are generally grown in loamy sand or sandy loam soils, whereas almonds are grown in a wider variety of soil types and geographical regions.

Reducing Nematode Populations

The worst place to plant a new orchard, from the standpoint of nematode control, is an old orchard or vineyard site. Nematode populations can occur as deep in the soil as tree roots have penetrated—tree roots in the soil die slowly and offer food and some protection from fumigation to nematodes for many years after the tree has been removed.

Whenever possible, plant new almond orchards in land that has previously been planted in nonwoody crops. The soil won't contain as many almond nematode pests as soil in an old orchard or vineyard site. In addition, fleshy crop roots rot more quickly than woody ones, leaving the nematodes unprotected in the soil. If you do plant in an old orchard or vineyard site, properly prepare the soil with fallowing and preplant fumigation, and use resistant rootstocks if root knot nematodes are present, or if the soil is conducive to their presence. Following is a suggested schedule for preparing an orchard site to reduce a nematode infestation:

Season	Action
Summer/Fall	Remove trees or vines, destroy residues, and deep cultivate to remove residual roots and break up cultivation pans or soil layering. Sample for nematodes.
Winter/spring	Fallow or plant grains.
Spring/summer	Level (if necessary), cultivate, and do other operations required for next year's planting. Dry the soil.
Late summer/ early fall	Rip the soil. You may need to apply an inch of water to break surface clods. Fumigate preferably in September or October.
Winter/spring	Observe waiting period on fumigant container label; plant young trees on resistant rootstock if root knot nematode is present.

Rootstock Selection

Selection of a particular rootstock is generally based upon a number of factors (page 19), one of which is resistance or tolerance to nematode attack. None of the more commonly used rootstocks is resistant to the root lesion, ring, or dagger nematode pests. Resistant rootstocks are available for root knot nematode, however, and are the primary line of defence against this pest.

If you prefer to use rootstocks that are susceptible to root knot, your success is most likely in finer-textured soils, including fine sandy loams, or in cooler regions such as the northern production areas of the state. Root knot resistant rootstocks include Nemaguard, some of the almond-Nemaguard hybrids, and Marianna 2624. These rootstocks are not resistant to the root lesion, ring, or dagger nematodes, however, and the use of any rootstock is limited by soil and horticultural considerations. Nemaguard, and to a certain extent almond-Nemaguard, rootstock is widely accepted for use in coarse-textured soils and in areas where root knot nematodes are prevalent. Root knot nematodes can penetrate resistant rootstocks but are unable to reproduce in them. This inhibition of reproduction is sufficient to greatly reduce or practically eliminate the development of all California species of root knot nematode in weed-free almond orchards. Occasionally, galling can occur on root knot resistant rootstocks, but these galls are uncommon and short-lived.

Lovell peach and especially almond rootstocks are susceptible to root knot nematodes and are not recommended for use on land infested with them unless extensive control measures are taken before planting, and even then their use is questionable. In addition, these two rootstocks are susceptible to attack by root lesion, ring, and dagger nematodes.

Cultural Practices

The speed with which a nematode infestation spreads depends upon how favorable the soil environment is to the nematode population and the manner in which nematodes were introduced into the orchard. If introduced with trees, nematodes may cause damage within the first year. Introduction with new trees, however, is rare in California because California nurseries must sell certified nematode-free stock. Nematodes are more commonly introduced into orchards via contaminated soil, equipment, or runoff water. In such cases, the nematode population needs time to become established and may never cause observable damage until you try to replant. Because nematodes can be transferred from one area to another, clean your equipment of soil and roots before moving from an infested to a noninfested orchard, and do not irrigate noninfested areas with runoff water from infested areas.

Intercropping. If you wish to intercrop during the first 2 years of orchard establishment, be careful about using broadleaf plants such as cotton, lettuce, and alfalfa. In general, these and most other broadleaf plants are good hosts for the root knot nematode, although some, such as alfalfa, will not show signs of galling. Infestations on these broadleaf crops in sandy soils can produce high nematode populations so rapidly that they will reduce growth of the almond tree in spite of the good "field resistance" available from Nemaguard rootstock. Similarly, because most broadleaf plants are good hosts for root knot nematodes, be careful in the choice of plants used in an orchard cover crop to prevent buildup of nematodes.

Cultivation and Drainage. Little is known about how to cultivate and irrigate to minimize nematode damage in an orchard. Current research indicates that root knot damage can be minimized on Lovell rootstocks by making water more frequently available to the root system. This can be accomplished by improving the water-holding capacity of the soil with soil amendments, such as manure or other organic matter, and by irrigating more frequently with sprinkler or drip irrigation systems. Neither practice reduces the nematode population, but the tree suffers less damage. These practices appear to be useful in minimizing only root knot damage, not that caused by ring or root lesion nematode.

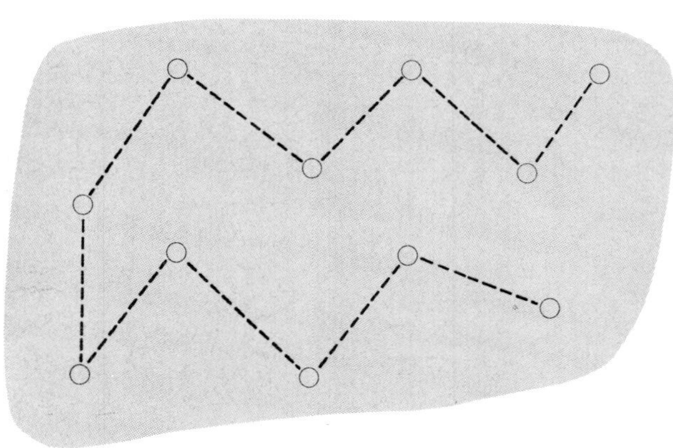

Figure 54. Recommended pattern for collecting soil core subsamples from a sampling block in a fallow field.

Fallowing or Planting Grains. Fallowing the land or planting cereal grains before you plant or replant an orchard can reduce nematode pest populations by removing their food source. The longer you fallow or plant in grains, the lower the population of nematodes parasitic to woody perennials will be.

Soil Samples

Soil samples taken from the planting site and examined in a professional laboratory will help you determine which nematode species are present. This information is important for soil fumigation and proper rootstock selection. High nematode populations frequently occur in soils that 1) were previously planted in woody crops; 2) were recently planted in broadleaf annuals; 3) are sandy or have a sandy subsurface; and 4) are in areas where bacterial canker is prevalent. If your soil falls into any of the above categories (or even if it does not), sampling before planting an orchard is recommended.

Some laboratories collect samples, or you may do it yourself. (See *Commercial Analytical Laboratories in California Available for Agricultural Testing,* listed under General References, or consult your farm advisor.) If you collect samples yourself, make sure the laboratory will be ready to process them as soon as they are received. Processing usually takes about 2 weeks. Keep the laboratory reports as part of the permanent orchard history.

To collect samples, visually divide the orchard site into sampling blocks that represent differences in soil texture, drainage patterns, or cropping history. Take a separate sample for each block so that each can be managed appropriately. A sample consists of about one quart of soil that is taken from a mixture of 5 to 20 subsamples. Nematodes are harder to detect when soil has been fallow or recently planted in nonhost crops such as cereal grains. In these situations take as many as 20 subsamples to increase the likelihood of detecting any nematodes present.

In a fallow field, collect subsamples from several locations within the sampling block. Use a pattern similar to the one in Figure 54. In an established orchard, collect separate subsamples from the soil around trees that show symptoms and from the soil around adjacent, healthy looking trees for comparison. Subsamples should include feeder roots when possible and be taken in frequently wetted zones at the edge of the tree canopy.

Sampling can be done at any time of the year. It is best to sample when the soil is still moist, preferably within a week after rainfall or irrigation. Take samples to a depth of 36 inches (90 cm). You can use a soil auger, a sampling tube (Viehmeyer, Oak Field, or the like), or a shovel (Figure 55). The amount of root collected with a sampling tube is much smaller than with a soil auger, but the soil volume of the sample is also reduced.

For each sample, collect the subsamples at various depths from 6 inches to 3 feet (15 cm to 1 m). Put the subsamples in a bucket, and mix them thoroughly; take one quart as a representative sample for the sampling block. Pour the soil and roots into a durable plastic bag or other moisture-proof container. Seal tightly and place the bags in the shade until you finish taking samples. Attach a label to each bag providing your name and address, location of the orchard, sample block, soil type or texture, cropping history (crops for last several years, current crop, anticipated crop), and notable symptoms; this information is critical for a meaningful analysis.

If you take more than one sample from each sampling block, it will improve the precision of the estimate and the information on nematode distribution in the field, but processing samples is costly. A useful rule of thumb is to use 1% of the cost of establishing an orchard for sampling nematodes before you plant.

If you want your samples analyzed for ring nematode, request that the nematodes be extracted by sugar flotation or centrifugation techniques. These procedures will also detect root knot nematode juveniles and root lesion nematodes. Any diagnosis of nematode samples, however, must distinguish *Pratylenchus vulnus*, which parasitizes the roots of almond trees, from a second species of root lesion nematode (*P. minyus*), which is common in grassy orchards but does not parasitize tree roots.

Send or deliver the samples to the laboratory as soon as possible. Ship them in a cardboard box insulated with newspaper or in summer, use a styrofoam ice chest. If any delay occurs, keep the samples in a cool place (40° to 55°F or 5° to 10°C).

Damage thresholds for the different nematode pest species in almonds have not been established. If your soil samples indicate that root knot, root lesion, or ring nematodes are present and the land to be planted has a crop history of woody perennials or broadleaf plants, you will need to fumigate your soil before planting.

Soil Fumigation

Nematode pests cannot be eradicated chemically over a large area. Nematode populations will gradually rebuild over time, and there are no postplant treatments available.

Preplant soil fumigation allows an almond tree time to develop a healthy root system that can ultimately withstand or tolerate some nematode damage when the populations rebound. If you replant blocks of trees on susceptible rootstock, fumigate the entire block. Trees planted on Nemaguard rootstock will also benefit from soil fumigation in replant situations. If only root knot nematode is present, you can save money by just fumigating the tree row or individual tree sites before planting trees with Nemaguard rootstock. If ring or root lesion nematodes are also present, fumigate the entire orchard area.

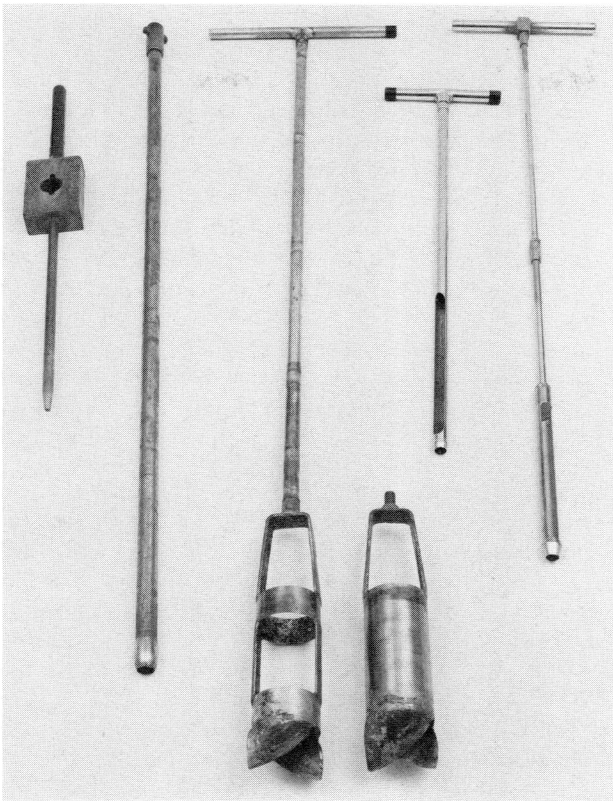

Figure 55. Tools for taking soil samples. The Veihmeyer tube (left) has a slotted slide hammer for driving the tube into the soil and removing it. Soil augers (center) are available with a variety of bits for different soil samples. Oakfield soil tubes (right) are usually the easiest to use, especially for samples down to about 2 feet (60 cm). These tools are useful in monitoring soil moisture and in sampling for nematodes.

The quantity of chemical applied is only one factor that determines the effectiveness of a soil fumigation. Equally important is the soil moisture content, soil texture, and soil temperature. Each of these must be properly evaluated before fumigation. Figure 56 lists application rates required for control of nematodes in the surface 2.5 and 5 feet (.75 and 1.5 m) of soil depending on soil conditions.

Fumigation of individual tree sites is more effective if each site is backhoed to 5 feet (1.5 m) and then the sides

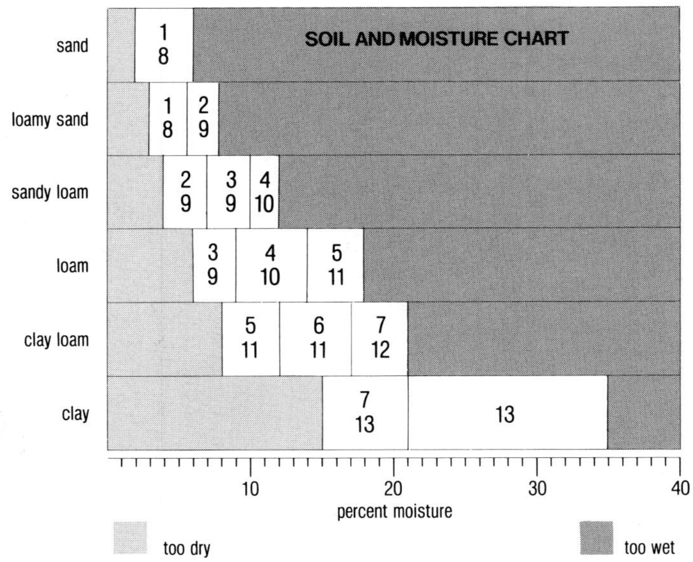

No. Code from Soil and Moisture Chart	Soil Temp. (°F)	Application Rates (lb/Ac)	
		1,3-D NEMATICIDES (92% 1,3-D)[1]	
		A	B
1	40 - 77	50 - 100	400
2	40 - 77	75 - 125	400
3	50 - 77	100 - 150	500
4	50 - 77	125 - 175	500
5	60 - 68	150 - 200	600
6	60 - 68	200 - 250	700
7	60 - 68	250 - 300	800
		TARPED METHYL BROMIDE (98% ai)[1,2]	
		C	D
8	40 - 77	200	400
9	40 - 77	250	400
10	40 - 77	300	500
11	40 - 77	350	600
12	50 - 77	400	700
13	50 - 77	500	800

[1]A—Controls soil pests, such as nematodes outside roots, throughout the surface 2.5 feet of soil.

B—Controls pests or nematodes in smaller (less than 2 in. diameter) roots throughout the surface 5 feet of soil.

C—Controls pests or nematodes in smaller (less than 2 in. diameter) roots throughout the surface 5 feet of soil.

D—Eradicative treatment to control nematode virus vectors throughout the surface 5 feet of soil.

[2]Additional chloropicrin may be present.

Figure 56. Application rates for preplant fumigation to control nematodes. Use the highest application rate recommended for the soil conditions within the profile. For example, if a soil has a loamy sand surface layer with 5% soil moisture and a subsurface layer of loam with 10% moisture, fumigate with the higher rate given for loam soil. A permit is required for the use of both 1,3-D nematicides and methyl bromide.

of the hole are caved in before fumigation. Because the soil will subside following planting, place extra soil on the tree site during the hole-filling and fumigation.

Old roots larger than 1/2 inch (1.25 cm) in diameter and tree stumps harbor nematodes and protect them from fumigation, so remove old roots and stumps. Do the same if you are planting individual trees within an established orchard, and do not irrigate the site before fumigating.

The best time to fumigate is usually from September until mid-November, but this period can vary regionally, and greater flexibility is possible in sandiest soils. Less than optimum soil conditions result in faster return of nematodes, and the difference can mean from 6 months to 6 years of nematode protection.

Soil Moisture. Correct soil moisture levels are important for successful fumigation. Determine soil moisture by taking samples at locations throughout the orchard. Be sure to sample in all areas of the orchard where soil textural variations occur. At each sampling location, take one-pint samples 1, 2, 3, 4 and 5 feet deep. Weigh each sample, dry it at 221°F (105°C) for 24 hours, and then weigh it again. Use the following equation to determine the percent moisture of each sample. Be sure to subtract the weight of the sample container from each weight determination.

$$\frac{\text{wt. of soil before drying} - \text{wt. of soil after drying}}{\text{wt. of soil after drying}} \times 100 = \text{percent moisture (on a dry weight basis)}$$

If soil moisture is too low, the liquid fumigant becomes highly adsorbed to the soil particles, reducing the effectiveness of the fumigant. Too much moisture also hinders effective fumigation by preventing fumigant movement. In addition, variations in the percent moisture within the profile can influence the movement of the fumigant. For example, if the moisture content exceeds 12% in a subsurface layer of sandy loam soil located at the three foot depth, the fumigant would not penetrate this layer (Figure 56). Unless the moisture content in the sandy loam layer was reduced before fumigation, only the top 3 feet would receive an adequate amount of fumigant for effective nematode control. Sandy soils are generally easier to fumigate than fine-textured clay soil because they are easier to dry out and prepare.

Soil Texture. Soil texture describes the size of soil particles. A coarse-textured soil contains a high percentage of sand and has large pores through which the fumigant disperses quickly. Fine-textured soils, on the other hand, have a high proportion of clay and silt, and the size of the pores is much smaller; such soils are capable of holding greater percentages of water. Fumigants move much slower and are more readily diluted in fine-textured soils.

Determine the texture of the soil at the site to be planted. Fumigants vary in their ability to move through finer-textured soils. The finer the soil texture, the higher the rate of application required for the fumigant to penetrate to the 5-foot depth (Figure 56).

Most soils are not uniform, but have textural variation within the profile. Restrictive layers such as hard pans, clay pans, and abrupt boundaries between different soil textures (Figure 57) may prevent adequate penetration of the fumigant. Soils with a shallow restrictive layer may require backhoeing. In addition, fumigants will not penetrate clods of soil or roots larger than 1/2 inch (1.25 cm) in diameter if they are within 6 inches (15 cm) of the soil surface; eliminate these from the surface before treatment if possible.

Soil Temperature. Make sure that the soil temperature is right for the material you are using. The best time to fumigate is in fall before the soil gets wet and cold. Check the soil temperature with a thermometer at a depth of 12 inches. Fumigants will not volatilize or penetrate roots if soil temperatures are too low; 1,3-dichloropropene* (Telone II) can be effective down to about 50°F (10°C), and methyl bromide* down to 40°F (5°C). Above a soil temperature of 77°F, the fumigants degrade and volatilize too quickly.

*Restricted material; permit from agricultural commissioner required for possession or use.

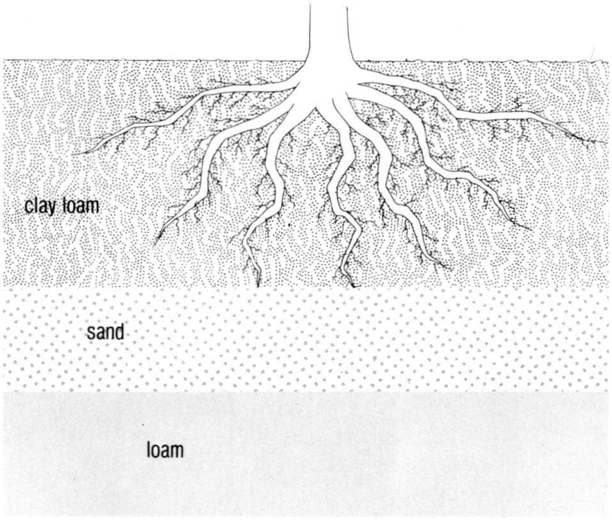

Figure 57. Abrupt boundaries between different soil textures act like pans, even though they may not be dense.

Application. The top several inches of soil are the most difficult to fumigate. Cultipacking, rolling, or dragging the soil immediately after fumigation can provide an adequate seal for 1,3-D fumigants. With gases, such as methyl bromide, cover the treated area with a plastic tarpaulin immediately after fumigation.

Some soil fumigants are toxic to plants and, depending on the dosage, must be applied several weeks or months before trees are planted. The pesticide label will indicate the potential phytotoxicity of the soil fumigant to the almond tree. Follow the specific directions on the fumigant label for placement and depth of chisels, and observe the suggested waiting period before replanting.

Weeds

Weed management in an almond orchard should reduce the negative impact of weeds on almond trees, prevent the buildup of hard-to-control weeds, and have the orchard floor free of plant debris at harvest. Permanent cover crops or sod cultures may not be suitable in an almond orchard because they make harvesting difficult. However, plants growing on the orchard floor at other times of the year can be advantageous. Their root systems, especially those of annual grasses, can penetrate plow pans and improve water percolation in many soils. Plants can provide a firmer soil surface for better year round access in the orchard. They also help reduce soil erosion as well as dust problems that create favorable conditions for mites. Finally, plants may provide refuge for beneficial insects and mites.

If not properly managed, however, weeds can create a variety of problems in the orchard. They can compete with almond trees for water, nutrients and sunlight, especially in newly planted orchards. Badly infested young orchards may take longer to come into economical production. Weeds are less competitive after 3 or 4 years when the orchard has become established, but many weeds, especially perennials such as bermudagrass, can still lower orchard productivity in other ways.

Weedy orchards may enhance the activities of other pests by providing shelter or overwintering sites for some species of pest mites and for insects that may vector almond diseases. Weeds can also serve as alternate hosts for nematodes or pathogens that damage almonds. For instance, the pathogens that cause almond leaf scorch, yellow bud mosaic, and green fruit rot may survive the winter on certain weeds that do not express any disease symptoms. Also, weeds left growing around the base of trees can provide shelter for mice as well as cause trunks to remain wet and increase the potential for crown rot. When the vegetation dries, these weeds can also create a fire hazard.

If weeds are not kept mowed during almond bloom, they can increase the chance of frost damage by blocking the warming influence of the soil: weedy orchards are slightly cooler than orchards with bare ground. In addition, weeds such as fiddleneck, wild radish, and mustards that flower at the same time as almond trees compete with the tree for pollination by bees. To avoid this competition, mow the ground cover just before bloom. In flood and furrow irrigated orchards, weeds may cause problems by restricting the distribution of irrigation water. At harvest, weeds growing around the trunks restrict the shaker operator's view and may contribute to trunk and limb injury. If not properly managed, weeds can also increase the amount of trash on the orchard floor and hinder the sweeping and pickup operations at harvest possibly causing part of the crop to be left in the orchards.

Growth Habits

All plants found growing in an almond orchard can be grouped as annuals, biennials, or perennials. Annual plants germinate, grow, flower, and produce seeds in one season. The cycle may be completed in several months at any time of the year depending on the weed species. Most annuals are classified generally as winter annuals or summer annuals. In California's mild climate, certain annuals may behave as biennials or shortlived perennials—for example, bristly oxtongue and cheeseweed (little mallow). Biennials complete their life cycle in two growing seasons, producing vegetative parts in the first and flowers and seeds in the second season. Perennials live 3 years or longer, often dying back in winter but regrowing the following spring from underground portions of the plants.

Management Guidelines

A plant's status as a weed is often closely related to the type of life cycle it has. For example, winter annuals are the least troublesome for growers because ample moisture is available in winter. Winter annuals can be managed by cultivating or mowing in early spring or by using herbicides. Summer annuals, biennials, and perennials, however, require more intensive management because they compete with trees for water and nutrients during the dry summer months and may interfere with irrigation and harvesting.

Perennials are generally the most difficult to control because most have extensive stolons, rhizomes, bulbs, tubers or tap roots, which will generate new growth after mowing or cultivation. The best time to control perennials is before the orchard is planted. If seedlings are found in young orchards, they should be controlled before they form a network of rhizomes and stolons. If established perennials are present, treat them with herbicides before they produce seeds.

Site Preparation

Weed management starts before the orchard is planted. In spring, survey the site for the weed species present and then disc the weeds under. At the same time, level any irregularities in the grade. Proper land preparation before planting, or at least before establishing the permanent irrigation system, is important for optimum performance and safety of herbicides used in the orchard.

Perennials on the site, such as johnsongrass or bermudagrass, are easier and less expensive to control before you plant the trees. Established johnsongrass and bermudagrass can be destroyed by repeated discing in the summer; the disced and exposed rhizomes and stolons will dehydrate. Seedlings can be controlled with herbicides before or after the orchard is planted. After discing two or three times in the summer, treat the perennials in early fall with glyphosate when they are still actively growing and flowering, and disc 10 to 20 days later to expose the root system to drying. If new growth is observed the following spring, spot treat with glyphosate. Before planting the trees in spring, a preemergence herbicide, such as trifluralin, can be incorporated over the entire site or into 4- to 6-foot strips where the trees will be planted. Napropamide and oryzalin can be applied after planting. Young almond trees are less susceptible to trifluralin, napropamide, oxyfluorfen, and oryzalin than to other preemergence or preplant herbicides. If field bindweed is present, trifluralin and oryzalin will help control the seedlings and reduce established plants. When planting the trees, place untreated soil directly around the roots, and then cover with a surface layer of treated soil. If you preplant fumigate for nematodes, however, be sure the soil you place directly around the tree roots has been fumigated but not treated with herbicides.

The spring preemergence application will usually control germinating seedlings of susceptible weeds throughout the summer. You may need to spot or broadcast treat surviving annuals and perennials with contact herbicides, depending on the extent of preplant control. Unless you plan to till the orchard after the trees are planted, apply a preemergence herbicide in late fall, just before the first rains, to control the winter weeds that you noted in your spring survey of the orchard site. Thereafter, apply an annual fall preemergence treatment to the tree row if you plan to use strip or basal weed control (weed control around the base of the tree); or treat the entire orchard floor if you plan total reliance on herbicides. Apply postemergence herbicides when monitoring indicates a need. In areas with heavy soils or high rainfall, splitting the preemergence application into two applications can prolong the control (see *Preemergence Herbicides*, page 139). Make the first treatment in fall and the second in late winter. When established weeds are present, both applications can be combined with a postemergence herbicide.

Management in Established Orchards

Weed management programs in established almond orchards may rely primarily on tillage (Total Cultivation), mowing (Total Mowing), or herbicides (Total Reliance on Herbicides); they may combine mowing or cultivating with the use of herbicides in strips down the tree rows (Strip Weed Control) or localized use of herbicides near the tree trunks (Basal Weed Control). The choice depends on your irrigation method, soil type, potential for frost, potential for erosion, personal preference, economic, and other considerations.

No matter what management strategy you choose for your orchard, make every effort to prevent introduction of undesirable weeds and eliminate conditions that may enhance development of weed problems. To avoid spreading weeds, keep irrigation ditches weed free. Also, clean your cultivating and harvesting equipment to avoid spreading underground portions of perennial weeds and weed seeds. Drain areas with standing water because they provide ideal conditions for dallisgrass, sprangletop, and curly dock. By letting the top 2 to 4 inches (5 to 10 cm) of soil dry completely between irrigations, you can discourage weed growth, especially the establishment of seedlings. Also, herbicides will degrade rapidly in soils saturated with water for long periods of time, resulting in poor weed control.

Biological control of weeds in orchards is limited. Puncturevine stem and seed weevils can help reduce the number of puncturevines in unsprayed orchards, but they are more successful in controlling puncturevine in undisturbed locations, such as roadsides. Purslane sawfly larvae feed inside leaves of purslane, where they reduce plant vigor and seed size. They also have a greater impact if the plant is growing in an undisturbed location.

Total Cultivation. Total or clean cultivation (tillage) keeps the orchard free of weed growth during most of the year. Often a newly planted orchard is tilled for the first year or two to eliminate established weeds and is then converted to nontillage. Cultivation may be the most practical way to manage your orchard floor if you irrigate with furrows or contour checks, which must be removed and replaced periodically. Cultivation in combination with herbicides is also a good method for controlling a heavy growth of perennial weeds, such as bermudagrass or johnsongrass. Chemically control perennials with a systemic herbicide, following label instructions. Do it before cultivation, however; otherwise, underground portions of the plant will be spread by the machinery, increasing the infestation. Another approach is to disc the perennials first to cut rhizomes into smaller pieces, irrigate the orchards so that the rhizomes begin to sprout, and then treat with glyphosate. Because the herbicide translocates faster in short rhizome segments than longer ones, this practice makes the weed easier to control chemically.

To carry out a cultivation program, disc or harrow the orchard floor once during late winter or early spring to turn under the winter ground crop, and four or more times during the summer, depending on the number of irrigations and amount of weed growth. You will probably have to weed around the trunks by hand several times a year, especially if your orchard is young. One way to prepare weed-free orchards for harvest is to disc and roll or float the orchard floor, and use an herbicide just before or in the last irrigation.

Weeds allowed to grow during winter will prevent soil erosion and improve soil structure. Where spring frosts are a problem, however, a clean, smooth, moist orchard floor can provide a slightly warmer orchard than mowing. In summer, shallow cultivations to control weeds can help to conserve water in areas where water is limited.

Total cultivation requires discing in both directions with heavy equipment that tends to compact the soil, decreasing water penetration. Decreased water penetration will increase the amount of runoff water in orchards planted on slopes and lead to soil erosion problems. In addition, discing may result in mechanical injury to the lower trunk, making trees susceptible to crown diseases. Discing also cuts feeder roots in the top 6 inches of soil; thus, the tree cannot use the rich supply of nutrients, oxygen, and water in this area. Do not disc any deeper than you need to. Cultivation can increase a weed infestation if perennials are not chemically controlled before discing. Cultivation generates dust, which may increase mite problems, so keep the number of times the soil is tilled to a minimum.

Total Mowing. To manage the orchard floor with total mowing, mow the annual winter ground cover instead of discing it into the soil. Mowing can either be done mechanically or chemically. Chemical mowing involves using low rates of a nonselective postemergence herbicide to maintain low growth of the ground cover. This method is still experimental, and the exact rate and timing of chemical applications have not been determined.

Winter ground cover is usually a volunteer cover provided by annual plants. If the volunteer ground cover is not adequate, it may be necessary to plant an annual fall-seeded cover crop, such as wheat, cereal rye, or barley at 50 to 80 pounds per acre, to start the program. For information on plants to use in cover crops, see *Cover Crops in California Orchards and Vineyards* listed under Weeds in References.

Begin mowing the ground cover in early February, before frost is a concern, and thereafter as necessary until harvest. Generally the plants should be mowed before they reach 6 to 8 inches (12 to 20 cm) or before seed heads form on tall growing weeds. Keeping the cover this height will allow desirable, low growing winter annuals, such as redmaids, chickweed, filaree, and annual bluegrass, to reseed themselves. To avoid injuring the tree trunks with the mowing equipment, weed around the trunks by hand or with herbicides (see basal weed control below). Following the last irrigation before harvest, flail the ground cover as close to the ground as possible without disturbing the soil surface. The objective is to have the orchard floor free of all weed growth by the time harvest begins. If the orchard is mowed frequently enough, the clippings will have sufficient time to decompose and will not interfere with harvest. Also, less trash will be trucked to the hullers.

Total mowing offers certain advantages. This system uses lighter equipment that is less expensive and easier to operate than the total cultivation system; it reduces soil compaction; and it makes the orchard accessible during wet conditions. Roots of cover plants help maintain good water penetration. Tree roots can grow closer to the surface to make use of water, oxygen, and nutrients in the top soil. A ground cover also reduces dust.

On the other hand, there are drawbacks with total mowing. While the cooling effect of the ground cover can be advantageous in areas with hot summers, it can increase frost hazards somewhat. You can minimize this problem by keeping the ground cover closely mowed in late winter and early spring if weather permits. Maintaining a ground cover requires more water and nutrients than a clean-cultivated orchard. Under mowing, perennials, such as bermudagrass, field bindweed, and dandelions, tend to become dominant in the weed population because they grow longer and compete better for nutrients and water than annuals do when mowed. Nutsedge is also favored by mowing in the lighter soils of the San Joaquin Valley. Finally, total mowing does not allow the use of contour checks for irrigation.

Strip Weed Control. Keeping weeds away from trunks without injuring trees is difficult in total mowing or total cultivation programs. You can avoid this by using strip weed control to maintain a 2- to 6-foot-wide strip within the tree row weed free with herbicides and mowing or discing weeds between the rows.

Use this system in both young and established orchards and with drip, sprinkler, or flood irrigation with permanent berms established down the tree row. Strip weed control uses less herbicide than does total reliance on herbicides. It also saves time and energy in tilled and mowed orchards because it eliminates the need for cross-discing and mowing. Strip application of herbicides reduces travel through the orchard with tillage equipment by 40% to 50%. It also prevents injuring trees and spreading diseases because the equipment does not come close to the tree trunks. By keeping the tree row free of weeds, there is less competition for nutrients and water, and tree trunks are kept dry. Strip weed control also offers all the advantages listed under total mowing.

When practicing strip weed control, watch for perennials in the mowed strip, on treated berms, and at the orchard's borders. Consider spot treating them with a postemergence herbicide. Once perennials become established, they grow vigorously in the absence of competition from annual weeds in the herbicide treated strips. For a detailed description of strip weed control, see *Non-tillage and Strip Weed Control in Almond Orchards* listed under Weeds in References.

Basal Weed Control. Herbicides in a basal weed control program are applied to weeds in a 4- to 8-square foot area around the base of each tree instead of down the entire tree row. Basal weed control requires less herbicides than strip weed control, but more mowing or tilling. Like strip weed control, it eliminates the need for mowing or tilling close to the trees, greatly reducing the possibility of trunk injury. It has the same advantages and disadvantages of strip weed control, but this method is not suitable for orchards with permanent berms or checks because it requires two-way mowing or tilling. In young orchards, use preemergence herbicides such as oryzalin, oxyfluorfen, and napropamide, which are relatively safe around young trees.

Total Reliance on Herbicides. The orchard floor can be kept free of weed growth if you spray the entire orchard floor with herbicides. Total reliance on herbicides requires fewer trips with equipment through the orchard than any of the other methods. Perennials may be easier to control with this method because the seedlings are easier to spot on bare ground. In areas where the average yearly rainfall is 11 inches or less, total reliance on herbicides can greatly reduce weed competition for water. Although this method is convenient, initially it may be the most costly.

Certain disadvantages are associated with total reliance on herbicides. No single herbicide will control all annual weeds found infesting almond orchards. Combinations and sequential herbicide applications are required to maintain the orchard floor free of unwanted vegetation. In orchards planted on slopes, soil erosion can be substantial. With some soils, orchards kept weed free with herbicides may be less accessible when wet than an orchard with a mowed ground cover. Soil compaction and the development of a thin, silty surface layer, which impedes water penetration, also may be a problem on certain soils. Light, shallow cultivation, or scratching the soil surface, may be needed as a remedy.

Most preemergence herbicides do not control established perennials, and these perennials will spread rapidly in the absence of annuals. This is particularly true of nutsedge and bermudagrass, which are also tolerant to most postemergence herbicides. If certain preemergence herbicides are used repeatedly, resistant species may thrive in the absence of competition. For example, simazine eliminates most broadleaf annuals, so that perennials

	PREEMERGENCE									POSTEMERGENCE				
	OXYFLUORFEN (Goal)	SIMAZINE (Princep)	DINOSEB 1/ (Premerge, Sinox PG)	NORFLURAZON (Solicam)	NAPROPAMIDE (Devrinol)	ORYZALIN (Surflan)	TRIFLURALIN (Treflan)	DICHLOBENIL (Norosac)	EPTC (Eptame)	GLYPHOSATE (Roundup)	DINOSEB 1/ (Dinitro)	MSMA 2/ (Bueno 6)	PARAQUAT 1/	WEED OIL
ANNUAL GRASSES														
Annual bluegrass	P	P		C	C	C	C	P	C	C			C	C
Barnyardgrass				P	C	C	C	P	C	C			P	C
Bearded sprangletop				P	C	C	C	P	C	C		P	P	C
Large crabgrass				C	C	C	C	P	C	C		C	C	C
Fall panicum				P	C	C	C			C			P	C
Johnsongrass seedling				C	C	C	C	P	C	C		C	C	C
Wild barley	C	P		C	C	C	C	P	C	C			C	C
Wild oats	C	P		C	C	P	P	P	P	C			C	C
Witchgrass	P			P	C	C	C		C	C			C	P
ANNUAL BROADLEAVES														
Cheeseweed (mallow)	C	P		P	P			P	P	P				P
Chickweed		C	C	C	C	C	C	C	C	C	C		C	C
Clovers	C	C			P			C		P			P	C
Common groundsel	C	P	P	C	C			C	C	C	P		C	C
Common lambsquarters	C	C	C	C	C	C	C	C	C	C	C		P	C
Fiddleneck	C	C	C	C	C	C	C	C	C	C	C		C	C
Filaree	C		P	C	P		C	P	P	P	P		P	C
Flax-leaved fleabane	P	C	P	P				P	C	C	P		C	C
Horseweed	P	C	P	P				P	C	C	P		C	C
Knotweed	C	C	P	C	C	C	C	P	C	C	P		P	C
Mustards	C	C	C	C	P			P	P	C	C		C	C
Pigweeds	C	P	C	P	C	C	C	C	C	C	C		C	C
Puncturevine	C			C	P	P	C	P		C		P	C	C
Purslane	C	C	P	C	C	C	C	C	C	C	P		C	C
Redmaids	C	C	C	C		C	C	C	C	C	C		C	C
Spurge	P		P	C		P		C	P	C	P		C	C
Wild radish	C	C	P	C	P			P	ND	C	P		P	C
PERENNIALS														
Bermudagrass				P		P	P			C				
Common dandelion								P		P				
Curlydock	P			P	ND	ND	ND	ND	ND	P				
Dallisgrass				ND		ND	ND			C		C		
Field bindweed						P	C	P		P		P		
Johnsongrass				P		P	P			C		C		
Nutsedges				P	P			P	P	P		P		

1/ Permit required.

2/ Registered for nonbearing orchards only.

(blank) not controlled | (P) partially controlled | (C) controlled | ND no data

Figure 58. Weed control provided by herbicides currently registered for use in almonds. This information is derived from field experience and is tentative; changes may be necessary as new information becomes available.

and resistant annual grasses, such as crabgrass and witchgrass, can build up and require additional postemergence management.

Use of Herbicides

Whether you apply herbicides to the entire orchard floor, to the tree row, or just to the area at the base of the tree, you will need to apply a preemergence (soil-applied, residual) herbicide annually. Apply long-lasting preemergence herbicides in the fall or early winter to kill weed seedlings as they germinate. Postemergence herbicides are applied when monitoring indicates the need.

Preemergence Herbicides. Preemergence herbicides must be incorporated mechanically or leached by water into the top 2 inches of soil where weed seeds germinate. Some herbicides must be incorporated immediately after application, whereas others will tolerate a short interval before incorporation; read the herbicide label to determine how soon the particular herbicide(s) must be incorporated after it is applied. In drip irrigated orchards or in surface irrigated orchards where trees are planted on berms, apply herbicides before a rain so that the material is carried into the soil. In orchards irrigated by sprinklers or flooding, most preemergence herbicides can be applied anytime before an irrigation. Examples of preemergence herbicides are oxyfluorfen (Goal), norflurazon (Solicam), napropamide (Devrinol), oryzalin (Surflan), and dichlobenil (Norosac). Simazine (Princep) is also a preemergence herbicide, but it must be used at very low rates in flood or sprinkler irrigated orchards, unless the soil is predominately fine textured or is high in organic matter. Preemergence herbicides are most effective on germinating weed seeds; at rates used in almonds, they generally do not control established weeds. Oxyfluorfen, however, has foliar activity on certain young, established weeds. Annual broadleaves and cheeseweed (a biennial) are more susceptible to oxyfluorfen as young plants than are annual grasses and perennials.

Preemergence herbicides can provide control from several weeks up to a year, depending on yearly rainfall, the solubility of the material, soil properties, frequency and method of irrigation, weed species, and dosage applied. Prolonged moist conditions during heavy winter rains or around low-volume emitters promote breakdown and leaching of herbicides. Splitting a preemergence treatment into two applications can prolong control, particularly in areas with heavy rainfall, in orchards with sandy soils, in orchards that were treated early in the fall, or in orchards with a heavy growth of summer annuals. Treatment can be split by using one-half to two-thirds the amount of chemical in the fall and the remainder the following spring.

In some instances, herbicides can be used at rates lower than specified on the label for season-long annual weed control. For example, in orchards where the trees are planted on berms and the average yearly rainfall is less than 11 inches, lower rates of herbicides may be effective because treated soil is often above surface-run irrigation water and not subject to the same degree of breakdown and leaching. Where berms remain dry during irrigation, weed seed germination is reduced, which further reduces weed problems.

A given dosage of preemergence herbicide is more toxic in sandy soils or soils low in organic matter than in soils high in clay or organic matter. Herbicides also leach from sandy soils more readily than from clay soils. Thus, in orchards with sandy soils, splitting the herbicide treatment gives longer residual control and is safer for trees.

Because preemergence herbicides can persist in soil from a few months to a year or more, discontinue using them 1 or 2 years before removing an orchard. Where it is necessary to replace a tree, make sure to backfill with untreated soil around the roots of the young tree.

Postemergence Herbicides. Postemergence (foliar-applied) herbicides are used on young, established weeds. They do not remain effective for very long in the soil and are applied when monitoring indicates that the preemergence treatment was not adequate or that perennials are present. Apply postemergence herbicides in the fall with the preemergence herbicide if weeds have emerged. Otherwise, apply them later in the season as a follow-up to the preemergence treatment. If you manage weeds by total cultivation or total reliance on herbicides, use postemergence herbicides to spot treat perennials.

The two types of postemergence herbicides can be distinguished by their mode of action. *Contact herbicides* kill those parts of the plant that are actually sprayed; good coverage and wetting are essential. Examples are weed oils, used alone or fortified with dinoseb; emulsions of water, weed oil, and dinoseb; or paraquat. A single spray kills susceptible annual weeds; retreatment is necessary if regenerating perennials are present or if annual weeds reestablish themselves from seeds. Contact herbicides are most effective when applied to weed seedlings or young weeds because it is easier to get good coverage, and less material is needed. The second type, *translocated or systemic herbicides,* does not require thorough coverage because the material is transported from the sprayed part to other plant parts, such as roots and rhizomes. Foliar-applied, translocated materials, such as glyphosate, 2,4-D, and MSMA, are therefore more effective in killing perennials.

Choose herbicides and application rates according to weed species present, soil type, irrigation method, and age of trees. No single material registered for almonds will control all weed species; combinations or sequential applications of different herbicides may provide better control than one compound alone in many situations. Follow

ORCHARD LOCATION ______________________ **HERBICIDE(S)** ______________________

DATE ______________________ **APPLICATION** ______________________

COMMENTS ______________________

	NOV ______		FEB ______		MAY ______	
	TREATED	UNTREATED	TREATED	UNTREATED	TREATED	UNTREATED
ANNUAL GRASSES						
annual bluegrass						
barnyardgrass						
crabgrass						
fall panicum						
sprangletop						
wild barley						
wild oats						
witchgrass						
ANNUAL BROADLEAVES						
cheeseweed (mallow)						
chickweed						
clovers						
filaree						
fiddleneck						
flax-leaf fleabane						
groundsel						
horseweed						
knotweed						
lambsquarters						
mustards						
pigweeds						
puncturevine						
purslane						
wild radish						
PERENNIALS						
bermudagrass						
curly dock						
dallisgrass						
field bindweed						
johnsongrass						
nutsedge						
dandelion						

all label precautions carefully. Figure 58 shows the approximate control provided by herbicides currently registered for almonds. For details of registration status and use of individual herbicides, contact your farm advisor, PCA, or county agricultural commissioner.

To apply herbicides properly, calibrate carefully and check the functioning of spray equipment; recommended procedures are described in *Calibration of Herbicide Sprayers* listed under Weeds in References.

Monitoring

Monitoring is essential for choosing the correct preemergence herbicide or combination of preemergence herbicides and deciding whether a postemergence herbicide is needed. Monitor your orchard for weeds three times a year: November, February, and May. In November after the first rains have begun, look for winter annual seedlings in the tree row to determine if the preemergence treatment was adequate. Also, check the ground cover for perennial seedlings. In February you will be able to record the full spectrum of grown winter weeds, as well as monitor for perennials. If a few winter weeds were not controlled by the preemergence treatment, note them for possible changes in next season's treatment. This frequently occurs if preemergence herbicides used in the tree row are ineffective against the dominant weeds in the orchard. If weed growth is heavy, you may need to consider a postemergence treatment. By May or June, summer annuals will have germinated. Monitoring at this time will reveal if the preemergence treatment was effective against summer annuals and what species of perennials are present.

If you rely on tillage for weed control, monitor before you cultivate, and treat perennials with a systemic postemergence herbicide to prevent spreading underground portions of the plant.

Records are an important tool in weed management. Weed survey information collected over several years is valuable in identifying changes in weed populations and in planning herbicide programs. Record your monitoring observations; a sample format is given in Figure 59. Pay special attention to perennials; sketch a map of the orchard and mark where they occur. Keep all records as part of the permanent orchard history.

Weed records can be stored on the computer. Develop a form similar to Figure 58 recording the weed species you find in individual orchards, the herbicides

◀ **Figure 59. Keep records of weed species in your orchard on a form such as this. Monitor three times a year. Use a scale from 1 to 5 to indicate level of infestation: 1 = very few weeds; 2 = light infestation; 3 = moderate infestation; 4 = heavy infestation; 5 = very heavy infestation. If you have treated and untreated areas, such as in strip weed control, record the weed species in each area separately.**

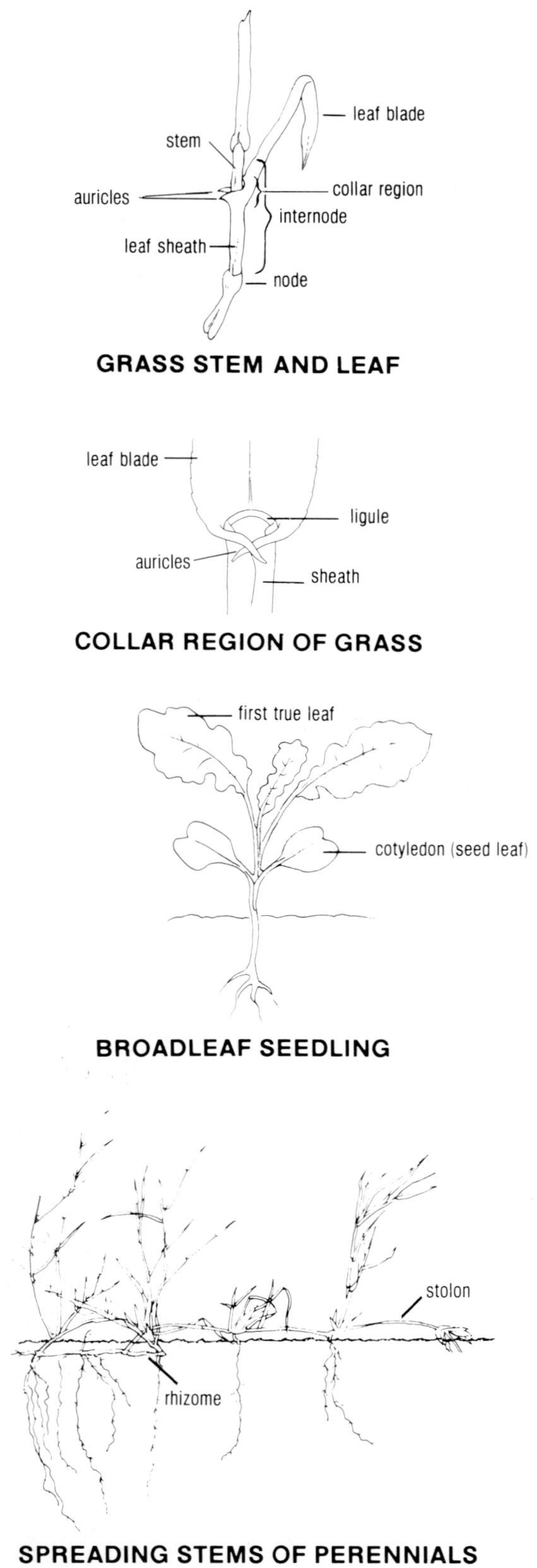

Figure 60. Vegetative parts commonly used in identification of weeds.

used, and whether or not control was achieved. Weed species can also be recorded by year or, for perennials, by location in the orchard. Knowing which weed species are present and their susceptibility to registered herbicides, will enable you to select more effective herbicides the following season.

Identifying Plants. Plants can be classified as *broadleaves, grasses,* or *sedges.* At emergence, *broadleaved* weeds have two seed leaves, often of characteristic shape, color, and texture. The seed leaves and usually the first true leaves differ from later leaves. *Grass* seedlings are difficult to identify; the collar region is usually the most distinguishing feature. Perennial grasses have characteristic rhizomes or stolons (Figure 60). *Sedges* superficially resemble grasses, but most sedges that grow in orchards can often be distinguished by their three-sided (cross sections are triangular) stems and closed leaf sheath. Nutsedges (nutgrass) are perennial sedges that have tubers or nutlets on the rhizomes. Umbrella sedge is sometimes confused with nutsedge, but it does not produce nutlets.

Management is easier if you identify and control weeds in the seedling stage. The following descriptions and photos should help you identify major species found in almond orchards. A few necessary technical terms are illustrated in Figure 60 and explained in the glossary. In most instances, you have to use a combination of characteristics to identify a species. Look at several young plants; the characteristics may be more or less pronounced in individual specimens, depending on their growing conditions. If you encounter weed species not covered in the chapter, check with the *Growers' Weed Identification Handbook* listed under Weeds in References or your farm advisor.

A

B

Major Weed Species

Grasses—Winter Annuals

Annual Bluegrass *Poa annua*

Annual bluegrass provides an excellent ground cover because it is low growing and dies early in the summer when the topsoil dries out. It can easily be distinguished from other grasses by its typical leaf tip and blade.

A. The mature annual bluegrass plant grows as dense, low-spreading tufts, 3 to 12 inches (8 to 30 cm) tall, and often roots at the lower nodes. The most distinguishing feature of the mature plant and the seedling is the blunt leaf tip shaped like the bow of a boat. The leaf blade is often crinkled at midsection.

Wild Oat *Avena fatua*

Wild oat is troublesome in orchards because its high growth around tree trunks holds moisture, making the crown susceptible to diseases, keeping the orchard from warming during frost periods, and providing shelter and protection to meadow mice. Because it produces large seeds that are capable of germinating at depths below the effective zone of most preemergence herbicides, it can be a difficult grass to control. Use napropamide in the pre-

C

emergence treatment, or use postemergence grass herbicides on the young plants. Wild oat dries up and dies in midsummer. The seedlings of wild oat and wild barley look very similar. Wild oat, however, retains the seed from which it is growing for a long time. Mature plants of the two species are easily distinguished by their different flower heads.

B. The mature wild oat plant is sturdy, 1 to 4 feet (30 to 120 cm) tall, with several stems arising from the base.

C. The collar region of wild oat has a tall, pointed ligule with toothed margins and no auricles (right); also shown are wheat (middle) and wild barley (left).

Wild Barley *Hordeum leporinum*

Wild barley is less bothersome than wild oat because it is much shorter and dries up earlier in the year. The mature plant grows as short, spreading tufts, 6 to 24 inches (15 to 60 cm) high. The flower head is a dense, bristly spike with erect awns; at maturity, the head breaks apart into several pieces. Wild barley has characteristic narrow, long auricles, which clasp the stem (Figure C); the auricles, however, may not be very pronounced in the young plant.

Grasses—Summer Annuals

Barnyardgrass *Echinochloa crus-galli*

Several varieties of barnyardgrass or watergrass grow in California; they vary mainly in the appearance of the flower head. Barnyardgrass grows in dense patches and is difficult to control by nonchemical means because it produces a tremendous number of seeds. It is competition sensitive but will proliferate in the absence of other plants. It is the only major summer weedy grass without a ligule and auricle.

D. The mature barnyardgrass plant is stout, grows upright, and varies from 6 inches to 6 feet (15 cm to about 2 m) high. It often roots at the lower nodes, forming dense tufts. The leaf sheath is flattened and often reddish tinged at the soil line. Lower spikes of the flower head are spaced apart; top ones are crowded together. Flower groups are barbed with short, stiff hairs or long bristles, depending on variety.

E. The collar region of barnyardgrass has no ligule and no auricles.

Bearded Sprangletop *Leptochloa fascicularis*

Bearded sprangletop is a summer annual commonly found in wet places throughout the interior valleys and is usually associated with alkaline soils. It thrives in wet places, such as ends of water runs and around stand pipes, where herbicides degrade quickly or are leached below the germination site. Provide good drainage in those areas to discourage growth.

F. Bearded sprangletop has no auricle, but it has a long, thin, delicate ligule.

G. The mature, bearded sprangletop plant is coarse, grows upright and forms large tufts 12 to 40 inches (30 cm to 1 m) tall. Flower heads are long, erect, well branched, and straw colored at maturity.

D

E

F

G

Large Crabgrass *Digitaria sanguinalis*

Large crabgrass, also known as hairy crabgrass, is widely distributed throughout orchards and vineyards, especially those in the San Joaquin Valley. It roots deeply at the nodes and is difficult to remove physically once it is established. Crabgrass is not controlled by simazine, but is easy to control with oryzalin, norflurazon, and napropamide.

H. The flowering stalk of large crabgrass resembles bermudagrass, but slender flowering branches usually rise from the stalk separately; in bermudagrass, the branches all rise from the same point.

Fall Panicum *Panicum dichotomiflorum*

Fall panicum, also known as smooth witchgrass, is a common weed in cultivated orchards, moist ground, and waste places. It is frequently found in orchards where only simazine is applied in the preemergence treatment. Fall panicum has thick, flattened stems that may either be erect or, more commonly, widespread and freely branching. Leaves are rolled in the budshoot, and the leaf blades are smooth, dull above, and glossy below with a prominent mid-vein. The collar is broad and continuous, and no auricles are present. The ligule is a dense fringe of hairs. Fall panicum grows in a clump and can be a problem at harvest if it remains in the orchard. In addition, its seed heads can clog up harvest equipment.

Witchgrass *Panicum capillare*

Witchgrass is prevalent in orchards, vineyards and irrigated crops growing in sandy soils. It is also known by the names tumbleweed grass, ticklegrass and witches' hair. It has a tough, wiry stem that frequently causes difficulty at harvest by interfering with nut pick up and contributing to the accumulation of trash in the harvested crop.

I. The mature witchgrass plant is bushy and branched from the base. Stems, sheaths, and leaf blades are covered with long, coarse hairs with swollen bases. Large, widely branched flower heads (see photo) become brittle at maturity and often break off and blow around in the wind.

Grasses—Perennials

Johnsongrass *Sorghum halepense*

Johnsongrass is one of the most troublesome perennial grasses in California. It commonly grows in annual crops, orchards, and along ditchbanks. The young plant may resemble barnyardgrass except barnyardgrass does not have a ligule. It does not form stolons but reproduces from rhizomes and seeds. Seeds may be dormant for many years and are frequently spread by irrigation water. Rhizomes are often spread by cultivation. Cultivating to expose and dry out the rhizomes will control johnsongrass before an orchard is planted if the soil is kept dry. In established orchards, johnsongrass is best controlled with herbicides.

J. The mature johnsongrass plant grows in stout, spreading and very leafy patches, which may be up to 6 to 7 feet (2 m) tall. The leaves have a prominent whitish mid-vein, which snaps readily when folded over. The flower head is large, open, well branched and often reddish tinged.

K. Underground stems (rhizomes) of johnsongrass are thick, fleshy, and segmented. Roots and shoots can rise from each segment.

L. The ligule in johnsongrass consists of a fringe of dense, fine hairs.

Dallisgrass *Paspalum dilatatum*

Dallisgrass is a perennial bunchgrass that is occasionally troublesome in orchards. Like johnsongrass, it does not produce stolons, but reproduces from rhizomes and seeds. It is not a good plant to have in a ground cover because it may be allelopathic and inhibit the growth of other plants. Also, its rapid growth enables it to quickly invade the tree row, where it is highly competitive for water and nutrients. Like bermudagrass, it tends to become dominant in mowed areas because mowing stimulates seed set. It also grows in areas with standing water. The plants are heavy seed producers, and the seeds are easily moved about by water or machinery. Because it grows in a clump, dallisgrass can also be a problem at harvest by interfering with nut pick up. Like bermudagrass and johnsongrass, it can be controlled by cultivation before an orchard is planted, but it must be chemically controlled in established orchards. Preemergence herbicides are effective only against seedlings; established plants often need repeated applications of a postemergence herbicide (glyphosate) for adequate control.

M. The ligule in dallisgrass is firm and membranous with a few spreading hairs at the margins and no auricles.

N. The mature dallisgrass plant forms coarse, loose bunches, 1 to 4 feet (0.3 to 1.2 m) high. The flower head consists of three to six spikes, that arise apart on the stem and are often drooping. The leaf sheath is somewhat flattened; at the base, it is very hairy, often tinged red and usually inflated. The rhizomes of dallisgrass are fairly short with short internodes that appear as concentric rings. Dallisgrass plants often grow prostrate in orchards that are mowed.

Bermudagrass *Cynodon dactylon*

Bermudagrass is commonly used as a turfgrass, but it is a troublesome weed in orchards. Bermudagrass is a perennial that reproduces from rhizomes, stolons and seeds. It frequently becomes a problem in mowed orchards because mowing increases the amount of light that the stolons receive, thus stimulating their growth. Like johnsongrass, bermudagrass can be controlled by cultivation before an orchard is planted if the soil is dry. Roots, rhizomes, and stolons will die if they are exposed to the sun. Cultivate the land several times before planting, and do not irrigate. In established orchards, control bermudagrass with herbicides.

O. The mature bermudagrass plant forms dense mats with spreading and branching stolons. Erect stems rise from axils of short leaves on stolons; these stems are 4 to 18 inches (10 to 45 cm) tall and have longer

H

I

J

K

L

M

N

O

leaves than the stolons do. The flower head consists of three to seven slender spikes radiating from one point at the tip of the stem.

Nutsedge *Cyperus* spp.

Nutsedges closely resemble grasses but may be distinguished from grasses by their triangular stem. Another commonly used name for this plant is nutgrass. Nutsedge is not controlled well with most chemicals, and its nutlets are readily spread by cultivation. It is especially troublesome in orchards that are managed totally by herbicides because it quickly spreads in the absence of competitors. To control nutsedge it is most important to prevent formation of the nutlets. Do not allow the plants to develop beyond the 6- to 7-leaf stage, which is when nutlets begin to form. MSMA in nonbearing orchards and glyphosate in bearing orchards, applied two or three times during the growing season, can provide partial control.

P. Mature nutsedge plants are erect, 1 to 3 feet high (30 to 90 cm). The stem terminates in a flowering head. The flower has several unequal linear leaves just below the head that resemble the ribs of an umbrella. The flower cluster is yellowish brown and flat.

Q. Nutsedges have extensive root systems that contain tubers (nutlets). Nutlets serve as food storage reservoirs and are also capable of sprouting. Nutsedge produces seeds, but the principal source of new growth and dissemination are the nutlets, which are abundantly produced and are responsible for the weed's extraordinary persistence.

P

Q

R

S

Broadleaves—Winter Annuals

Mustards *Brassica* spp.

A number of different mustard species are established in California. Most commonly found in orchards are black mustard (*B. nigra*), wild mustard (*B. kaber*) and wild turnip or common yellow mustard (*B. campestris*). Mustards have frequently been used in ground covers because their taproots help open soil. They are not a good ground cover plant for almond orchards, however, because they bloom at the same time as the almond trees do. Also, mustards growing at the base of the tree hold moisture around the tree trunk, making the crown susceptible to diseases. Because mustards are large plants with a great volume of foliage, they can be difficult to disc. Mustards also compete with the trees for water and are particularly competitive in areas where the average yearly rainfall is less than 11 inches (27.5 cm).

R. All mustards look similar in the seedling stage. The seedlings also resemble those of wild radish. The very young seedling may even be mistaken for bindweed because both species have broad seed leaves with a deep notch at the tip. The first true leaves, however, are distinctive—deeply toothed and often hairy.

S. Mustards are characterized by rather large lower leaves, usually more or less deeply cut into irregular lobes, and by distinctive seed pods. The mature plant is erect and about 2 to 6 feet (0.5 to 2 m) tall. Leaves grow alternately on the stem. Many yellow flowers rise at the tip of the stalk. They have four petals, which spread in the form of a cross. The flower stalk extends during the blooming season, and maturing seed pods have long beaks. Redmaids, *Calandrinia ciliata*, can be seen growing in the foreground of this photo.

Wild Radish *Raphanus sativus*

Wild radish is an erect annual, or occasionally biennial, that is widely distributed throughout California. Like the mustards, it is not a good ground cover plant for an almond orchard because it blooms at the same time as the almond trees.

The mature plant, which can range from 2 to 5 feet (0.6 to 1.5 m) high, branches at the top. Leaves are covered with short, stiff hairs and vary in size and shape. Flowers range from white to purple with streaks of different colors and are sometimes tinged with yellow.

Redmaids *Calandrinia ciliata* var. *menziesii*

Redmaids are native to the western U.S. and generally are distributed at lower altitudes in California. Their showy flowers make redmaids conspicuous in early spring. It is a good plant to have in the orchard ground cover because, like filaree, it matures early and doesn't create trash problems at harvest. It also is easy to control with herbicides in the tree row.

S. (foreground). Mature plants are from 6 inches to a foot (15 to 30 cm) tall and much branched from the base. The succulent leaves are narrow, 3/4 to 2 inches (2 to 5 cm) long and alternate. Flowers are showy, crimson, with five petals notched at their tips. The lower part of each petal is light, giving the center of the flower a white appearance.

Chickweed *Stellaria media*

Chickweed is a low-growing weed often seen spreading out in extensive mats. Chickweed is a good plant for a ground cover or cover crop, but is usually combined with other plants such as bluegrass because it is not competitive enough to grow by itself. Although it is a winter annual, chickweed may survive the summer in shady, cool areas that offer sufficient moisture. With the onset of hot, dry days in the Central Valley, chickweed soon dries up.

T. Chickweed mats may cover a large area. Stems are trailing, weak, and slender, with a line of hairs down the side and sometimes rooting at nodes that are in contact with the soil. The mature leaves are ovate, 1/4 to 1 inch (0.5 to 2.5 cm) long and opposite on the stem. Chickweed flowers are small but showy, with five deeply cut white petals. They are borne singly on long slender stalks arising at the base of the leafstalk.

Whitestem, Redstem, and Broadleaf Filaree *Erodium moschatum, E. cicutarium,* and *E. botrys*

These are good plants to have in the orchard ground cover because they are low growing and not too competitive. As moisture becomes scarce in summer, they mature, dry up, and decompose by harvest.

U. Filarees are similar in growth habits and flower and seed characteristics. All three can either grow erect or have spreading prostrate stems 3 to 12 inches (7.5 to 36 cm) long. In the early stages, leaves form a rosette close to the ground.

V. The three species can be distinguished from each other by their leaf shape. Leaves of whitestem filaree are toothed (left), leaves of redstem filaree are deeply cleft or dissected (right), and leaves of broadleaf filaree are lobed (not pictured).

W. Filaree flowers resemble those of *Geranium* and are rose purple. The fruit is long and needle-like. At maturity the fruit separates into five parts; each having a tail that coils spirally as the seed matures and dries.

T

U

V

W

X

Y

Z

A

B

Broadleaves—Summer Annuals

Prostrate Knotweed *Polygonum aviculare*

Prostrate knotweed is troublesome in orchards because its tough, wiry stems can get entangled in the sweeping machinery at harvest. It thrives well in most orchards, even in poor and compacted soil. Knotweed belongs to the buckwheat family; plants in this family have a torn sheath encircling the stem at the leaf base and usually swollen nodes.

X. Seed leaves of prostrate knotweed are long, narrow, and rounded at the tip. They are light green with white streaks or blotches. The true leaves are much broader, emerging from an encircling, membranous sheath at the leaf base.

Y. The mature knotweed plant grows prostrate to erect with slender, tough stems, which are extensively branched, giving the plant a zigzag appearance.

Puncturevine *Tribulus terrestris*

Puncturevine has a dense cover of hairs and an extensive root system, which make it very drought resistant. Its characteristic spiny seed burrs are easily disseminated by one's shoes and vehicle tires. Tractors frequently deposit the seeds on the periphery of the ground cover strip, where they germinate. The plant then grows over into the strip and creates a problem at harvest. The mature plant is somewhat difficult to control with herbicides, although seedlings are not. The seed, however, can germinate at depths greater than the effective zone of some preemergence herbicides, making control difficult. A stem and seed weevil attacks puncturevine, but it is most successful in controlling plants growing in undisturbed locations, such as roadsides.

Z. The puncturevine seedling has thick, elongate, and brittle seed leaves that are green above and grayish underneath; they are creased along the prominent midvein. True leaves consist of 8 to 12 leaflets.

A. The mature puncturevine plant grows prostrate on open ground but almost erect in dense vegetation. The yellow flowers are borne singly in the leaf axils. The fruit consists of a cluster of five spiny nutlets or burrs; it breaks apart at maturity, and each burr has two spines.

Common Purslane *Portulaca oleracea*

Purslane thrives under both wet and dry conditions. It grows rapidly and if the preemergence treatment breaks down in summer, it can be a problem at harvest. Nuts that come into contact with its moist, succulent growth will retain moisture, especially if the nuts are stacked and not turned. Purslane prefers loose, nutrient-rich, sandy soil. A sawfly feeds on the leaves and reduces plant vigor but is not effective in controlling the weed in disturbed locations, such as an orchard. Horse purslane (*Trianthema portulacastrum*) looks similar to common purslane but has broader leaves and purple flowers.

B. The mature common purslane plant is prostrate or erect, 6 to 12 inches (15 to 30 cm) tall. The plant branches at the base and along the stems. Leaves are succulent and often reddish tinged. Small yellow flowers are borne singly or in clusters of two to three in stem axils or at the tip of stems. They usually open only on sunny mornings.

Horseweed *Conyza canadensis*

Horseweed, also called mare's tail, is widespread in California orchards, as well as roadsides, cultivated fields, and waste places. An increasing problem, it is somewhat tolerant of all the preemergence herbicides available for almonds except simazine. Simazine, however, has restrictions on use in light soils and soils low in organic matter. If horseweed is allowed to grow in the tree row, the tall woody stems will hinder harvest operations. If it is kept mowed throughout the summer, however, it should not cause problems at harvest.

C. If undisturbed, the mature plant grows erect, only branching toward the upper part, reaching heights of 1 to 10 feet tall. If mowed, the mature plant is shorter and branched. Frequently, mowed plants are confused with flax-leaf fleabane, but horseweed has dark green leaves while flax-leaf fleabane has gray-green leaves. The hairy leaves of horseweed grow alternately on the stem. Small, flowering heads with yellow centers are borne on the branched stems at the top.

D. Seed leaves of horseweed are oval and narrow towards the stalk. First true leaves are also oval, but later leaves are more narrow; all are covered with short hairs. Flax-leaved fleabane seedlings closely resemble those of horseweed, and the two plants are difficult to distinguish from each other in the seedling stage.

Flax-Leaved Fleabane *Conyza bonariensis*

Flax-leaved fleabane, often found growing in the same areas as horseweed, can be distinguished from horseweed by its more abundant, smaller, gray green foliage. Horseweed grows taller, if allowed to grow undisturbed, and has broader leaf blades. As seedlings, however, the two plants are difficult to distinguish from each other. Flax-leaf fleabane may be more tolerant of preemergence herbicides than horseweed. It can withstand several mowings and still produce seed. It is not a good plant to have in a ground cover because its hard stem does not degrade easily and can cause harvest problems. In addition, it can interfere with moving sprinkler and drip irrigation lines.

E. The mature flax-leaved fleabane plant is 1 1/2 to 3 feet (0.5 to 1 m) tall, branched, and has very narrow leaves that grow in an alternate manner. The upper portion of the branches contains many flowering stalks. Flowers are small, dull white, and not showy. The seed heads are slightly larger than those of horseweed.

C

D

E

F

G

H

Broadleaves—Biennial

Little Mallow *Malva parviflora*

In California's mild climate, little mallow or cheeseweed usually overwinters and is therefore sometimes considered a biennial. It thrives on nutrient-rich and sandy loam soils. Because of its tolerance to many herbicides, cheeseweed often becomes the predominant ground cover plant in orchards. The mature plant forms dense bushes that trail along the ground or grow upright, 1 to 4 feet (0.3 to 1.3 m) tall. It has a tough, woody stem that interferes with moving irrigation lines. When mowed, cheeseweed becomes a low-growing, branched, thick-stemmed plant that doesn't degrade well and causes problems at harvest. Several mallow species occur in California, but cheeseweed is the most common in orchards. Its flowers are white with a bluish or pinkish tinge and held in clusters in the leaf axils; the fruit resembles a miniature wheel of cheese.

F. Seed leaves (leaf on the right) of cheeseweed are distinctly heart shaped and tinged red. True leaves are roundish with wavy, shallow-toothed margins and a red spot at the leaf base.

I

Broadleaves—Perennials

Curly Dock *Rumex crispus*

Curly dock is one of the most common docks in California. As members of the buckwheat family, they are characterized by jointed stems, a membranous sheath at the leaf base, and usually swollen nodes. Curly dock usually grows in wet areas and is frequently associated with overwatering. It can be controlled by cultivation, but preemergence herbicides are not effective against its deep, heavy taproot. Curly dock can regrow from its thick, fleshy taproot when disced or mowed. Also, its heavy stem can interfere with harvest machinery.

G. The mature curly dock plant has stout stems, which are often reddish. It may be 2 to 5 feet (0.6 to 1.6 m) tall or low-growing when mowed. The long, loosely branched flower head is green but turns brownish red at maturity. In the fall, the stem dies back and a basal rosette of leaves is formed.

Field Bindweed *Convolvulus arvenis*

Field bindweed or perennial morningglory is a troublesome perennial because it is highly competitive, especially in young orchards. It competes for water and nutrients and may increase other pest problems, such as vertebrates, as well. It is difficult to control chemically after the seedling stage because it develops and reproduces from an extensive rootstock and produces abundant seeds; seeds may remain dormant for 40 to 60 years. Cultivation can propagate bindweed by spreading its rootstock. It prefers dry, warm, nutrient-rich, deep loams.

H. Seed leaves of field bindweed are broad and notched at the tip. Plants sprouting from rhizomes lack seed leaves. The first true leaves are heart shaped and deeply lobed at the base; later leaves are generally shaped like an arrowhead. Leaf stalks, or petioles, are grooved on the upper surface. Once young plants reach the 5-leaf stage, they begin to form their extensive root system and become very difficult to control.

I. The mature bindweed plant grows prostrate with slender stems, 3 feet or longer, or upward while entwining other plants. Flowers are funnel shaped, white to light red, and are borne singly on slender stalks in leaf axils.

Common Dandelion *Taraxacum officinale*

Dandelion is a widely distributed perennial weed. It is most troublesome in mowed orchards with heavy soils. Dandelion can reproduce from seed or can regrow from its strong, well-branched taproot. Not only is the plant difficult to control chemically, it also tolerates close mowing and can be spread by cultivation. The mature plant arises from a strong, deep taproot and has no visible stem. Leaves are clustered in a rosette at the base of the plant. The flower heads are borne singly on the tip of a 3- to 12-inch (7.5 to 30 cm) long hollow stalk, are bright yellow, 1 to 2 inches (2.5 to 5.0 cm) across, and consist of petallike ray flowers. The pappus on the seeds frequently clogs up tractor radiators.

Glossary

abdomen. the posterior of the three main body divisions of an insect.

abiotic disease. a disease caused by factors other than pathogens.

allelopathic. chemical inhibition by plants.

allowable depletion. the proportion of available water that can be used before irrigation is needed (Figure 13).

annual. a plant that normally completes its life cycle of seed germination, vegetative growth, reproduction, and death in a single year.

anticoagulant. a substance that prevents blood clotting, resulting in internal hemorrhaging; may be used as a rodenticide.

auricles. the earlike projections at the base of leaves of some grasses; used to identify species (Figure 60).

available water. the amount of water held in the soil that can be extracted by plants (Figure 13).

awn. a substantial hair or bristle that terminates a plant part.

axil. the upper angle between a leaf and the stem from which it is growing.

biofix. an identifiable event that signals when to begin degree-day accumulation.

biotic disease. a disease caused by a pathogen.

biotype. a strain of a species that has certain biological characters separating it from other individuals of that species.

blank. nut with no kernel—consists of only the collapsed pellicle (skin).

blight. a disease characterized by general and rapid killing of leaves, flowers, and branches.

bulb. an underground storage organ, composed chiefly of enlarged, fleshy leaf bases.

calcareous soils. soils containing high levels of calcium carbonate.

calibrate. to standardize or correct the measuring devices on instruments; to adjust nozzles on a spray rig properly.

cambium. thin layer of undifferentiated, actively growing tissue between phloem and xylem (Figure 2).

canker. a dead, discolored, and often sunken area (lesion) on a root, stem, or branch.

chorion. the outer membrane of an insect egg.

chlorophyll. the green pigment of plant cells, necessary for photosynthesis.

chlorosis. yellowing or bleaching of normally green plant tissue usually caused by the loss of chlorophyll.

cocoon. a sheath, usually mostly of silk, formed by an insect larva as a chamber for pupation.

collar region. in grasses, the region where the leaf blade and sheath meet; it is used in identifying species (Figure 60).

cotyledons. the first leaves of the embryo formed within the seed and present on seedlings immediately after germination; seed leaves (Figure 60).

crawler. the active first instar of a scale insect.

crochets. tiny hooks on the prolegs of caterpillars.

cultivar. a cultivated plant variety; synonymous with variety.

degree-day. a measurement unit that combines temperature and time, and is used in calculating growth rates.

dehiscence. opening naturally and regularly along lines of weakness; in fruits, opening along sutures to release seeds.

developmental threshold. the lowest temperature at which growth occurs in a given species.

diapause. a period of physiologically controlled dormancy in insects.

disease. any disturbance of a plant that interferes with its normal structure, function, or economic value.

dormancy. a state of inactivity or prolonged rest.

ectoparasite. a parasite that lives on the outside of its host.

endoparasite. a parasite that lives inside its host.

endosperm. the tissue containing stored food in a seed that surrounds the embryo and is eventually digested by the embryo as it grows (Figure 5).

embryo. the small plantlet within the seed; in almond, the embryo develops into the kernel (Figure 5).

epidermis. the outermost layer of cells on the bodies of animals or on plant surfaces.

evapotranspiration. the loss of water from soil by the combination of soil surface evaporation and transpiration by plants.

feeder roots. the youngest roots with root hairs, most important in absorption of water and minerals.

field capacity. the moisture level in soil following an irrigation and rapid drainage by gravity (Figure 13).

frass. a mixture of feces and food fragments produced by an insect in feeding.

girdle. to kill or damage a ring of bark tissue around a stem or root; such damage interrupts the transport of water and nutrients.

hibernaculum (hibernacula: pl.). a shelter occupied during the winter by an insect, notably peach twig borer.

immune. exempt from infection by a given pathogen.

indexing. testing a plant for a virus infection, usually by grafting tissue from it onto an indicator plant.

infection. the entry of a pathogen into a host and establishment of the pathogen as a parasite in the host.

inner bark. in older trees, the living part of the bark, comprised of phloem (Figure 2).

instar. the larval or nymph stage of an immature insect between successive molts.

larva. the immature stage, between egg and pupa, of an insect such as a moth, beetle, or fly; for nematodes, any life stage between the embryo and adult.

lesion. a well-defined area of diseased tissue, such as canker or leaf spot.

ligule. in many grasses, a short membranous projection on the inner side of the leaf blade at the junction where the leaf blade and leaf sheath meet (Figure 60).

metamorphosis. a change in form during development.

microorganism. an organism of microscopic or small size.

molt. in insects and other arthropods, the shedding of skin before entering another stage of growth.

mummy. unharvested nut remaining on the tree.

mutation. the abrupt appearance of a new, heritable characteristic as the result of a change in the genetic material of one individual cell.

mycelium. the vegetative body of a fungus, consisting of a mass of slender filaments called hyphae.

natural enemies. organisms that prey on or parasitize an organism (prey, host), thereby limiting the population of that prey or host.

necrosis. death of tissue, usually a well-defined part of a plant such as a leaf or part of a leaf or the areas inside a canker.

nodes. the leaf-bearing joints on plant stems (see Figure 60).

nucellus. in plants, the watery tissue composing the chief part of the young ovule in the flower and inside the seed during early development. It furnishes nutrients to the young embryo and is digested by the developing endosperm and embryo (Figure 5).

nymph. the immature stage of insects such as grasshoppers and aphids that hatch from eggs and gradually acquire adult form through a series of molts without passing through a pupal stage.

outer bark. in older trees, the exterior part of the bark that is dead and sometimes cracked (Figure 2).

oviposit. to lay or deposit eggs.

pappus. the modified calyx of flowers in the sunflower family; usually takes the form of bristles, scales or awns.

parasite. an organism that lives in or on the body of a larger living organism (its host); in this manual the term is also used to refer to insect parasitoids, which spend their immature stages on or within the body of a single host.

pathogen. any disease-producing organism.

peduncle. the stem of an individual flower or fruit.

pellicle. the covering (skin) that encloses the kernel; it is white during development but becomes brown at maturity.

perennials. plants which may live three or more seasons and flower at least twice.

petiole. the stalk connecting the leaf to a stem.

pheromone. a substance secreted by an organism to affect the behavior or development of other members of the same species; sex pheromones that attract the opposite sex for mating are used in monitoring certain insects.

phloem. the tissue of the inside layer of bark next to the cambium containing specialized cells, the most important being food-conducting tissue; in the aboveground parts of the tree phloem forms a thin, continous layer between the outer bark and the cambium and water-conducting vessels (xylem or wood); in roots, it is organized with the xylem in the central stele (Figure 2).

photosynthesis. the production of carbohydrates from carbon dioxide and water in the presence of chlorophyll, using light energy.

phytotoxic. causing injury to plants.

predator. any animal (including insects and mites) that kills other animals (prey) and feeds on them.

preemergence herbicide. herbicides applied before emergence of weeds.

prothorax. the anterior of the three thoracic segments of an insect.

pupa. the nonfeeding, inactive stage between larva and adult in insects with complete metamorphosis.

pustule. small blisterlike elevation of epidermis from which spores emerge.

resistant. able to tolerate conditions harmful to other individuals of the same species.

rhizome. a horizontal, underground shoot, especially one that forms roots at the nodes to produce new plants.

rootstock. the lower portion of a graft that makes up the crown and root system.

rosette. a cluster of leaves arranged in a compact circular pattern, often at a shoot tip or on a shortened stem.

scion. the portion above a graft that becomes the trunk, branch, and tree top; the cultivar or variety.

secondary infection. infection by microorganisms that enter the host through an injury caused previously by another pathogen.

sedges. a group of grasslike, herbaceous plants that, unlike grasses, have unjointed stems. Stems are usually solid and often triangular in cross section.

seed leaves. the first leaf (grasses) or first two leaves (broadleaf plants) on a seedling; cotyledons (Figure 60).

senescent. growing old; aging.

seta (setae: pl). a bristle

sheath. the part of a grass leaf that encloses the stem below the collar region (Figure 60).

soil profile. a vertical section of the soil through all its horizontal layers, extending into the parent material.

spiracle. an external opening of the system of ducts, or tracheae, that serves as a respiratory system in insects.

spore. a reproductive body produced by certain fungi and other organisms, capable of growing into a new individual under proper conditions.

sticktight. nut with hull firmly adhering to shell.

stolon. a trailing aboveground stem or shoot, often rooting at the nodes and forming new plants (Figure 60).

sucker. shoot arising from the trunk or rootstock.

suture. visible seam on hull.

systemic. distributed throughout a plant, usually via the vascular system.

taproot. the large primary root that grows vertically downward giving off small lateral roots.

thorax. the second of three major divisions in the body of an insect, and the one bearing the legs and wings.

tolerant. able to withstand the effects of a condition without suffering serious injury or death.

true leaf. any leaf produced after the cotyledons (Figure 60).

tuber. a much enlarged, fleshy underground stem.

vascular system. the system of plant tissues that conducts water, mineral nutrients, and products of photosynthesis through the plant, consisting of the xylem and phloem.

vector. an organism able to transport and transmit a pathogen to a host.

vegetative growth. growth of stems, roots, and leaves—but not flowers and fruits.

viroid. a portion of infectious nucleic acid, without the protein coat of a virus.

virulent. capable of causing a severe disease; strongly pathogenic.

virus. a small infectious agent, consisting only of nucleic acid and a protein coat, that can reproduce only within the living cells of a host.

wing. extension of the nut shell at the suture line; varies in size according to cultivar.

xylem. plant tissue that conducts water and nutrients from the roots up through the plant (Figure 2).

zonate. marked with zones or bands; belted; striped.

References

General

Almond Orchard Management. Publication 4092.*

Almond Production. Leaflet 2463.*

Commercial Almond Production in California: Information for Prospective Growers. Leaflet 2824.*

Commercial Analytical Laboratories in California Available for Agricultural Testing. Leaflet 3024.*

Degree-Days: The Calculation and Use of Heat Units in Pest Management. Leaflet 21373.*

Degree-Days in Relation to an Integrated Pest Management Program. Leaflet 21301.**

Horticulture

A Permanent Sprinkler System for Deciduous Orchards and Vineyards. Leaflet 2435.*

Basic Irrigation Scheduling. Leaflet 21199.*

Diagnosing Soil Physical Problems. Leaflet 2664.*

Drip Irrigation Management. Leaflet 21259.*

Grading Land for Surface Irrigation. Leaflet 2692.*

Honey Bees in Almond Pollination. Leaflet 2465.*

Irrigating Deciduous Orchards. Leaflet 21212.*

Managing Compacted and Layered Soils. Leaflet 2635.*

Managing and Modifying Problem Soils. Leaflet 2791.*

Measuring Irrigation Water. Leaflet 2956.*

Planting Layouts for Deciduous Orchards. Leaflet 2437.*

Questions and Answers about Tensiometers. Leaflet 2264.*

Soil and Plant Tissue Testing. Bulletin 1879.*

Soil Physical Environment and How It Affects Plant Growth. Leaflet 2280.*

Pesticides

A Guide to Controlling Almond Pests, Diseases and Micronutrient Deficiencies. Leaflet 21343.*

Calibration of Herbicide Sprayers. Leaflet 2710.*

Fumigation of Inhull Almonds on the Farm (Revised 1980). Available from county Extension offices or the Almond Board, P.O. Box 15920, Sacramento CA 95852.

How Much Chemical Do You Put in the Tank. Leaflet 2718.*

Pesticide Application and Safety Training Manual. Publication 4070.*

Pesticide Toxicities. Leaflet 21062.*

Reducing Pesticide Hazards to Honey Bees with Integrated Management Strategies. Leaflet 2883.*

Safe Handling of Agricultural Pesticides. Leaflet 2768.*

Toxicity of Pesticides and Other Agricultural Chemicals to Honey Bees—Laboratory Studies. Leaflet 2287.*

Vertebrates

Controlling Ground Squirrels Around Structures, Gardens, and Small Farms. Leaflet 21179.*

Guide to Vertebrate Pest Control Materials Registered in California. Leaflet 21226.*

Pocket Gopher Control with Mechanical Bait Applicator. Leaflet 2699.*

Vertebrate Pest Control Handbook. California Department of Food and Agriculture, Sacramento 1975.

Wildlife Pest Control Around Gardens and Homes. Publication 21385.*

Insects and Mites

Managing Mites in Almonds: An Integrated Approach. UCIPM Publication 1. 1984. Available from the Almond Board, P.O. Box 15920, Sacramento, CA 95852 or from the IPM Implementation Group, University of California, Davis, CA 95616

Monitoring Oriental Fruit Moth Development with Day-Degrees. Leaflet 7157.**

Diseases

Almond Disease Guide. Leaflet 2609.*

Plants Resistant or Susceptible to Verticillium Wilt. Leaflet 2703.*

*These publications are available from the University of California, Agricultural Sciences Publications; see page 2 for address. A free catalog lists many other titles on agricultural topics.

**Available from county Extension offices in California.

Nematodes

General Recommendations for Nematode Sampling. Leaflet 21234.*

Weeds

Calibration of Herbicide Sprayers. Leaflet 2710.*

Cover Crops in California Orchards and Vineyards. USDA Soil Conservation Service, Washington, D.C. 1983.

Growers Weed Identification Handbook. Publication 4030.*

Nontillage and Strip Weed Control in Almond Orchards. Leaflet 2770.*

Weed Control in Almond Orchards. Leaflet 21069.*

Suppliers

Box Traps (ground squirrels)
Joe Cook
11508 Keith Drive
Whittier, Ca 90806 (The Better Squirrel and Rat Trap)
(213) 695-6191

Live Traps
Tomahawk Live Trap Co.
P.O. Box 323
Tomahawk, WI 54487 (various sizes available)
(715) 453-3550